DESIGN FOR
RELIABILITY

ELECTRONICS HANDBOOK SERIES

Series Editor:
Jerry C. Whitaker
Technical Press
Morgan Hill, California

PUBLISHED TITLES

AC POWER SYSTEMS HANDBOOK, SECOND EDITION
Jerry C. Whitaker

THE COMMUNICATIONS FACILITY DESIGN HANDBOOK
Jerry C. Whitaker

THE ELECTRONIC PACKAGING HANDBOOK
Glenn R. Blackwell

POWER VACUUM TUBES HANDBOOK, SECOND EDITION
Jerry C. Whitaker

THERMAL DESIGN OF ELECTRONIC EQUIPMENT
Ralph Remsburg

THE RESOURCE HANDBOOK OF ELECTRONICS
Jerry C. Whitaker

MICROELECTRONICS
Jerry C. Whitaker

SEMICONDUCTOR DEVICES AND CIRCUITS
Jerry C. Whitaker

SIGNAL MEASUREMENT, ANALYSIS, AND TESTING
Jerry C. Whitaker

FORTHCOMING TITLES

ELECTRONIC SYSTEMS MAINTENANCE HANDBOOK
Jerry C. Whitaker

THE RF TRANSMISSION SYSTEMS HANDBOOK
Jerry C. Whitaker

DESIGN FOR RELIABILITY

EDITED BY
DANA CROWE & ALEC FEINBERG

CRC Press
Boca Raton   London   New York   Washington, D.C.

Library of Congress Cataloging-in-Publication Data

Design for reliability / managing editor, Dana Crowe; technical editor, Alec Feinberg; co-authors, Carl Bunis ... [et al.].
　　　　p.　cm.
　　Originally published: Lowell, MA : M/A-COM, c2000
　　Includes bibliographical references and index.
　　ISBN 0-8493-1111-X (alk. paper)
　　　1. Electronic apparatus and appliances—Design and construction—Quality control.
　　2. Reliability (Engineering) I. Crowe, Dana. II. Feinberg, Alec. III. Bunis, Carl.

TK7836 .D473 2001
620′.00452—dc21
　　　　　　　　　　　　　　　　　　　　　　　　　　　　　　　　　2001025115
　　　　　　　　　　　　　　　　　　　　　　　　　　　　　　　　　　　　CIP

Visit the CRC Press Web site at www.crcpress.com

© 2001 by CRC Press LLC

No claim to original U.S. Government works
International Standard Book Number 0-8493-1111-X
Library of Congress Card Number 2001025115
Printed in the United States of America 1 2 3 4 5 6 7 8 9 0
Printed on acid-free paper

PREFACE

To stay in business, you must develop and maintain a **Value Proposition** that is compelling. Whether you are a high-volume manufacturer of commercial products or a low-volume, high-value-added producer, over the last decade **Reliability** has become a critical aspect of that proposition.

Customers have come to expect that every product shipped will work the first time and every time. This is evidenced in the PPM, Six Sigma, and Cpk programs that we, as manufacturers, are expected to deliver. One needs to understand these measures are not just the turn-on quality of the product being considered, but the measure of that product's ability to survive the manufacturing cycle as well as meet end-customer use expectations.

If the concept of reliability is brought down to a more personal level, we make decisions that have the ability to financially affect a manufacturer in a profound way. If our last major purchase failed to meet our expectations or was constantly in the repair shop, the replacement for that purchase would undoubtedly be from a different manufacturer.

Companies function in a similar but somewhat more formal manner. When field returns data or in-house first past yield data continue to flag a component with an issue, that component is then designed out. Extending that philosophy, **Market Share** will be lost to an alternative manufacturer if a product fails in customer qualification, causing major development delays.

The successful implementation of a reliability system depends on a true concurrent engineering team focus. Developing the business infrastructure necessary to facilitate that process may require a considerable investment in personnel and equipment, but the return on investment will be immeasurable, producing more than its share of **Earnings Before Interest and Taxes (EBIT)**.

This text has been developed for both engineers and managers to provide a clear understanding of how **Design for Reliability**, the **DfR** concept, enhances the concurrent design cycle. The stage gate process provides the maximum level of benefit to the design while minimizing the cycle time impact to that process, producing the highest levels of product reliability possible.

This is the understanding that is behind a company's investment in the people, the process, and the tools to establish a capable, responsive, and innovative reliability program that positively impacts all phases of design, development, and production. Using this focus, you will learn how to go beyond solidifying a basic offering to the marketplace and create a true **Competitive Advantage**.

The goal is to bring a product to market using a concurrent engineering cycle that is focused on designing out and/or mitigating the potential **Physics of Failure Modes** prior to product release. At this point, you are truly developing reliable products to meet your customers' needs and creating your organization's **Value Proposition**.

ACKNOWLEDGMENTS

Design for Reliability is a revised and edited version of the M/A-COM Tyco Electronics *Design for Reliability* manual. This text was produced by a team of industry experts while employed with M/A-COM Tyco Electronics Engineering. I would like to thank the team of contributing authors, without whom this manual would not have been possible: Carl Bunis, Peter Ersland, and Alec Feinberg.

I would like to acknowledge and thank the management team at M/A-COM Tyco Electronics, Richard Hess and Gregory Stephens-North, for the opportunity to publish this text and for their ongoing support.

I would also like to thank the copy editors, Hei-Ruey Jen, Ruth Lowder, and Janice Sleeper for their efforts in producing this manual.

During the years of development, many have contributed to this manual. The list of those who have added so much to this effort is vast. It is for your help that I say a special thanks.

Regards,

Dana Crowe
Managing Editor/Co-Author

ABOUT THE EDITORS AND AUTHORS

Carl Bunis, Author, is a Senior Principal Engineer/Materials Scientist at M/A-COM/Tyco Electronics in Lowell, Massachusetts. He received his B.S. in Mechanical Engineering and M.S. in Materials Science and Engineering from Worcester Polytechnic Institute and has worked in reliability physics and failure analysis for over 13 years. He has extensive experience with diverse types of materials, including microelectronics, plating, soldering, aerospace, automotive and medical technologies. Mr. Bunis has published papers and given lectures and tutorials at local, national, and international conferences on root cause analysis, physics of failure, corrosion, intermetallic, alloy, phase determination by the employment of phase diagrams, and start-to-finish analysis.

Dana Crowe, Managing Editor and Author, is the Manager of Engineering and Technology for M/A-COM, Inc., Lowell, Massachusetts, and has over 18 years of industry experience. M/A-COM, Inc., an operating unit of Tyco Electronics, is a global business producing products ranging from GaAs material to RF components as well as full systems for the commercial wireless markets and defense industry. Mr. Crowe is responsible for growing M/A-COM's reliability program to a point where it is currently recognized as a world leader in product reliability by its customers. He has established the reliability technical staff, analysis laboratory, environmental laboratory, Intranet analysis database system, and the design for reliability methodology in use at M/A-COM. In addition, he is responsible for corporate mechanical engineering services, design drafting, documentation, product safety, design automation, and a centralized product data management system.

Peter Ersland, Author, has been a Senior Principal Engineer at M/A-COM/Tyco Electronics in Lowell, Massachusetts, for over 14 years and is currently responsible for investigating and assuring the end-of-life reliability of all M/A-COM semiconductor processes, including design and development of life test systems, design, execution, and analysis of reliability studies, and thorough electrical characterization of new semiconductor processes. Mr. Ersland is a member of the American Physical Society (APS) and of the Institute of Electrical and Electronics Engineers (IEEE). He is also the M/A-COM representative to JEDEC. He holds a B.S. degree in Physics from Gustavus Adolphus College, St. Peter, MN, and an M.S. in Physics from Mankato State University, Mankato, MN. Mr. Ersland has over 20 publications and presentations in the areas of GaAs reliability and RF testing.

Alec Feinberg, Technical Editor and Author, is Senior Principal Reliability Engineer at M/A-COM/Tyco Electronics in Lowell, Massachusetts. He received his M.S. and Ph.D. in Physics from Northeastern University. He has been working in the area of reliability physics for over 19 years and has previously worked at TASC and AT&T Bell Laboratories. He has experience in reliability of electronic systems, accelerated testing, parametric reliability analysis, semiconductor reliability of plastic ICs, hybrids and assemblies, thermodynamic reliability analysis electric vehicles, advance batteries, seismology, and aircraft corrosion. Dr. Feinberg has numerous publications and presentations and is a member of the IEEE.

LIST OF ACRONYMS

ARG	Accelerated Reliability Growth
C-SAM	Scanning Acoustic Microscopy
CDM	Charge Device Model
Cp, Cpk	Process Capability Indices
DART	Design Assessment Reliability Test
DfR	Design for Reliability
DMT	Design Maturity Testing
EDS	Energy Dispersive X-Ray Spectroscopy
ESD	Electrostatic Discharge
ESS	Environmental Stress Screening
EST	Environmental Stress Test
FET	Field Effect Transistor
FIB	Focused Ion Beam
FIT	Failure In Time (1 Failure per 10^9 Device Hours)
FM	Failure Mode
FMEA	Failure Modes and Effects Analysis
GaAs	Gallium Arsenide
HALT	Highly Accelerated Life Test
HAST	Highly Accelerated Stress Test
HBM	Human Body Model
HTOL	High-Temperature Operating Life
MESFET	Metal-Semiconductor Field Effect Transistor
MMIC	Monolithic Microwave Integrated Circuit
MTBF	Mean Time Between Failure
MTTF	Mean Time To Failure
MTTR	Mean Time To Repair
OEM	Original Equipment Manufacturer
SAM	Scanning Auger Microanalysis
SEM	Scanning Electron Microscopy

TABLE OF CONTENTS

Section I: The Stage Gate Process

Authors: Dana Crowe & Alec Feinberg

Section II: Supporting Stage Gate

Authors: Carl Bunis & Peter Ersland

Section III: Topics in Reliability

Authors: Dana Crowe & Alec Feinberg

THE STAGE GATE PROCESS

CHAPTER 1

Reliability Science

1.1 Introduction

Today's marketplace demands reliability. Meeting that challenge requires developing a reliability engineering team that supports the full design-development process. The reliability team performs three fundamental activities as shown in Figure 1.1: Design for Reliability, Reliability Verification, and Analytical Physics. These activities are the building blocks to a sound reliability program that sits on a foundation of concurrent engineering.

The first reliability science activity in support of product development is **Design for Reliability (DfR).** This starts in the Idea Phase of the product development cycle and continues through product obsolescence. Design for Reliability is used to affect the design for a positive product reliability improvement by utilizing physics-of-failure knowledge to design out potential problems. This process is interrelated with the two other building block activities, forming a coherent stage gate/phase design process.

The second activity is **Reliability Verification.** Here, verification studies and demonstration tests ensure meeting customers' reliability objectives. Reliability Verification takes place in two main forms: *Process Reliability* and *Design Maturity Testing*. Process Reliability focuses on the development of a fundamental understanding of a platform's inherent reliability and provides the foundation to develop a realistic accelerated design maturity test. Design Maturity Testing demonstrates that a product's failure rate and a customer's needs will be met when the product is exposed to demanding conditions. The third activity, **Analytical Physics,** is designed to collect knowledge about a product's physics-of-failure. Understanding the nature of how and why a product can fail is the key to designing and building a product that will meet our customers' expectations.

Figure 1.1
The product development building blocks

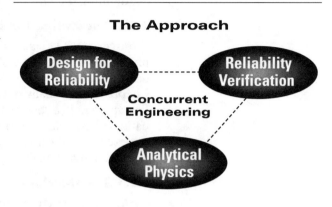

1.2 Reliability Design: "A Stage Gate Approach"

Reliability engineering with its three major activities supports a phased product development cycle called **stage gate.** The stage gate method is shown in Figure 1.2. The stage gate effort underpins product development, starting with product conception and continuing through final product obsolescence, including post-production. The stage gate method is essential in designing a reliable product capable of meeting customers' expectations. Ensuring that designs will meet customers' needs starts with an understanding of the full design requirements, environments hazardous to full product operating life, potential product use and misuse, total product cost goals, and reliability service life needs.

Figure 1.2
Stage gate process

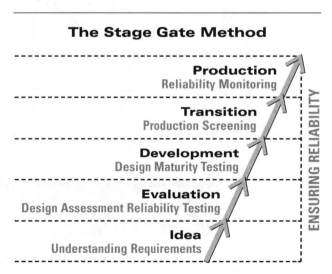

1.2.1 The Idea Phase

Stage gate 1: The concurrent engineering process of DfR activities begins with understanding customer requirements. Often these activities include the tools of Failure Modes and Effects Analysis (FMEA), product competitive benchmarking, and the reliability predictive modeling used to direct the design approach. The process actually starts in the Idea Phase stage and continues through the full product development cycle. These tools are deployed with the goal of reducing the risks associated with the first-pass success of a design's launch into the market. The first real impact on reliability occurs in this stage, for it is in the Idea Phase where the first concept of a design solution is selected, defining the ultimate reliability level that can be achieved.

1.2.2 The Evaluation Phase

Stage gate 2: Design risk-mitigation activities will occur during the evaluation phase. Risk-mitigation studies are performed to resolve uncertainties around the design approach. These *Design Assessment Reliability Test (DART)* studies (see Chapter 3) are usually not statistically sampled, but can investigate potential fatal flaws present in a chosen design. Reliability growth is the major focus of the evaluation phase where normally a 65 percent improvement in reliability can be achieved from the initial design point (see Figure 1.3). This is accomplished usually through test-analyze-and-fix activities in design risk-mitigation studies.

1.2.3 The Development Phase

Stage gate 3: The primary function of reliability engineering in this phase occurs toward the end of the development process. Here, *Design Maturity Testing (DMT)* is used to demonstrate and validate that a design will meet the expected operating-life requirements identified in stage gate 1. Design Maturity Testing is based on performing a physics-of-failure approach, knowing and understanding the physical issues within the design, and demonstrating that those issues will not impact the product within its useful life environment. This is accomplished by performing statistically significant failure-free accelerated life testing. A physics-of-failure approach is used in developing the Design Maturity Test. Tests in this phase are based on reliability science, historical information, process physics-of-failure studies, environmental product limits, and product environmental objectives. Typically, the statistically significant tests ensure that the product will meet its reliability objective at a 90 percent confidence level.

Figure 1.3
Growth reliability in stages

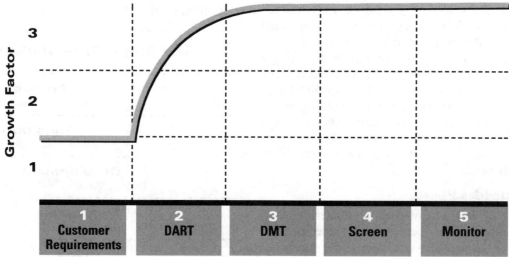

Accelerated Reliability Growth

Growth Factor

Stage Gate Processes

| 1 Customer Requirements | 2 DART | 3 DMT | 4 Screen | 5 Monitor |

Phase	Stage Gate	Task	Description
1	Idea	Understanding Customer Requirements	Concurrent engineering approach tools such as Failure Modes and Effects Analysis (FMEA) and competitive benchmarking are deployed to define the true design requirements.
2	Evaluation	Design Assessment Reliability Testing	Risk-mitigation studies and reliability growth efforts focus on finding and fixing failure modes in real time, concurrent with the design process, using the techniques of highly accelerated reliability growth, highly accelerated life testing (HALT) with Test-Analyze-and-Fix (TAAF), and others as necessary for the technology.
3	Development	Design Maturity Testing	Demonstrate that a design is reliably meeting the customer's expectations. Perform statistically significant accelerated tests, usually failure-free.
4	Transition	Production Screening	Ensure early production units are robust. Check for infant mortality problems.
5	Production	Reliability Monitoring	Ensure continual product reliability and quality to design obsolescence.

Table 1.1
Stage gate reliability

1.2.4 The Transition Phase

Stage gate 4: A reduction in reliability activities starts at this stage. The primary function of reliability engineering at this point is to help define the *proper screening* to prevent infant mortality failures from escaping to the customer. Defining the proper level of screening is not always simple. Failure mechanisms must be found. The correct screening technique to excite failure mechanisms must be identified and employed at the right level and duration to detect flaws without removing useful product life or inducing latent failures.

1.2.5 The Production Phase

Stage gate 5: The reliability engineering activities at this point are limited to defining the proper *reliability monitoring* process and techniques to ensure that the design continues to deliver the same performance over its lifetime. Reliability monitoring ensures process variations of the production cycle do not affect product reliability.

1.2.6 Defining How Much Is Enough

Reliability engineering engaged early in the design cycle and supporting the design to product obsolescence ensures a successful design release. Determining the level of reliability effort associated with product development, an assessment of the project's risk level is necessary. Often a design may be a spin-off of an existing design or a slight modification of a proven design. These products would not automatically dictate a full stage gate design process, using all the tools of reliability engineering. If a product is based on new technology and is revolutionary in nature, a high-level stage gate effort is to be performed. Because financial risk is linked to technology, it, too, is factored into an assessment. A balance and selection of these tools are needed to ensure that the highest levels of reliability are achieved, but not at a cost level that makes them prohibitive to the marketplace. The concept of risk is fully explored in Chapter 13. Table 1.1 describes the nature of the task associated with each phase of the design process.

1.3 Design for Reliability Tools

Designing a reliable product today is truly a concurrent engineering process. All design disciplines must be part of the product's development to ensure a robust design that meets a customer's needs. A reliability engineering approach with its series of tools can focus the design process. An overview of tools described here is shown in Figure 1.4.

The first major tool to be used is Failure Modes and Effects Analysis (FMEA)/Benchmarking. This is an important tool to ensure that reliability is integrated with product design (see Chapter 12). The FMEA tool can identify both specified and unspecified customer requirements for a design, how failure may occur, the severity of such failure, and the probability of the failure occurring. With these factors identified, we can focus the design process on the major issues of the product and its potential use environment. An FMEA provides the highest return for effort expended when concurrent engineering is properly applied.

Competitive Benchmarking is also important in assisting the DfR process. Such Benchmarking in the design process ensures that all important design aspects have been incorporated. Assessment of the competition is important, as they may similarly be assessing you. Leveraging all possible inputs available in today's challenging marketplace is part of best commercial practices that ensure state-of-the-art product performance, materials, packaging, and mechanical and electrical integrity, and identify cost issues.

Figure 1.4
Tools supporting Design for Reliability

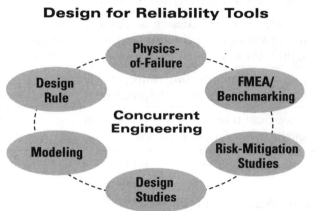

Design for Reliability Tools

Reliability predictive modeling is used to make initial product Mean Time Between Failure (MTBF)/Mean Time To Failure (MTTF) estimates. MTBF/MTTF estimates are important in understanding the feasibility of a design's capability of meeting the reliability goals needed to satisfy customer requirements. Also, such calculations direct and assist in the determination of design tradeoffs to ensure that the best design approach is taken. Predictive modeling activities continue through all stage gates to refine MTBF/MTTF estimates and ensure that a product will meet its reliability target. Although predictive modeling is known to have accuracy limitations relating to the models and database information available, excellent comparative studies still result.

Physics-of-Failure is critical to a reliability engineer's ability to affect the design process. In simple terms, Physics-of-Failure is an understanding of the physical properties of the materials, processes, and technologies used in the design and how those properties can interact with the life hazard conditions placed on the design during the product's full life cycle. The reliability engineer must understand the customer's use and misuse conditions and component/environment interactions to assist the design team in working around limitations inherent in the selected design approach.

Design studies, such as process reliability efforts, are and can be a full stage gate development process, and they are often associated with the development of a product. Design studies identify platform capabilities. They define a design's ability to meet the end-of-life requirements. They provide the necessary data to calculate an accurate MTBF/MTTF and give the necessary information to design an accurate zero-failure Design Maturity Testing platform to demonstrate the design's ability to meet the customer's needs.

1.4 Reliability Verification

Reliability Verification, the second building block activity, is often thought of as a group of tests, but Reliability Verification is much more. To perform Reliability Verification studies effectively, an extensive capability must be in place to simulate those environmental life hazard conditions placed on products and technologies by customers in an accelerated, compressed time period. Therefore, proper environmental tools are needed to stimulate the weaknesses of a design.

Figure 1.5
Reliability Verification building block

To support these stage gate environmental testing activities, a full in-house center must exist for environmental testing and evaluation. This testing facility requires equipment such as thermal shock chambers, accelerated life-test chambers, vibration systems, and humidity chambers to simulate operating conditions and to uncover weaknesses in a design. A full environmental test facility and the understanding of how to define and perform proper tests are critical to develop meaningful data from a planned study. Data acquisition techniques are a major component in understanding the complexities of a design while undergoing the extensive testing cycle of Reliability Verification and should not be overlooked. The techniques of verification have been discussed (see Figure 1.5).

Reliability Verification Techniques

1.5 Analytical Physics

The third building block activity is Analytical Physics, providing physics-of-failure data necessary to build robust technologies in today's demanding market. As technologies advance and products increase in sophistication, analysis capabilities must increase as well. Such advances are somewhat cyclical, currently improving by orders of magnitude about every six years.

To meet this challenge, an in-house world-class analysis facility is needed. This facility must be capable of performing product and process analyses, construction and reverse engineering, materials and physical analysis, and failure-mode investigative studies, including thermal and electrostatic discharge probing. The analysis facility allows for building upon past experience and historical information. Having the right information at the right time enables the design team to make correct and timely decisions for product designs. The key analysis tools shown in Figure 1.6 are fully described in Chapter 7.

Figure 1.6
Analytical Physics building block

Today's information technology of databases, intranets, and browsers has made it possible to put historical information online throughout a company's intranet. Real-time information is invaluable for the design team to address key questions and for others as the design's development process progresses. This insures that physics-of-failure information is utilized in the removal of failure modes. Linking the tools of analysis to a central database and providing direct access, via the intranet, provide the virtual co-location necessary to support engineering operations around the world.

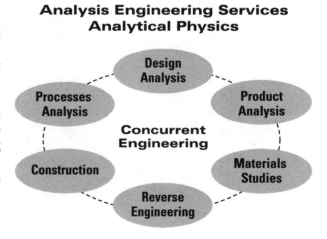

**Analysis Engineering Services
Analytical Physics**

1.6 The Goal Is Customer Satisfaction

To embark on an extensive program of reliability science and bring a company into the new-world marketplace demands an extensive commitment to the future. Oftentimes, the long-term combined fruit of this effort is not seen in real time. The process requires having to progressively build a world-class reliability operation. This consists of a fully dedicated staff of reliability engineers working on all key business drivers, to ensure that all products meet and exceed customers' expectations. The bottom-line result of the stage gate process is the production of products with dramatically low parts per million return rates. Designs that are correct the first time and released to the marketplace on time will delight customers.

In the next chapter, the details of reliability science will be reviewed in order to provide the reader with the tools necessary to start and implement a world-class reliability program.

CHAPTER 2

Understanding Customer Requirements

2.1 Introduction

Understanding the requirements of new technologies or products sounds like a very basic issue. In the past, especially in the defense industry, when a product met a customer's detailed specification, it was shipped. Ensuring that it met the customer's needs was not a requirement.

Today's world is constantly changing. Customers no longer set requirements in such detail. They rely on manufacturers to understand their needs and to anticipate what is required. This is a major shift in the thinking process of many manufacturers in the defense industry. Understanding a customer's requirements has become a major task.

The best way to understand a customer's requirements is by talking directly with them. Although many customers do not specify their requirements, they do have a good idea of what they expect from a product or service. A customer's requirements become clear when a manufacturer listens carefully to their needs. It is then that a manufacturer's challenge is to translate their customer's feelings and wishes into products.

Common sense plays a key role. For example, when manufacturing a plastic-packaged Integrated Circuit (IC) with leads that are soldered to the customer's circuit board, one requirement is that the soldered leads should provide reliable connections to the printed wiring board. A second requirement of the process is that the stress of the attachment process does not damage the device. These requirements are easy to understand. An area that is less clear is the misuse factor. Understanding how and why misuse occurs and how much to safeguard a design against misuse is another issue. A simple case that most people can relate to is a beeper dropped on the ground. The customer expects it to survive this fall. However, if it is dropped off a building, a customer does not expect it to survive. Understanding and defining such a distinction is a major factor in establishing a customer's expectations and, ultimately, in generating customer satisfaction.

Figure 2.1
Questions to ask in understanding customer requirements

As higher-value solutions are provided (for example, a subsystem rather than a simple component), it becomes even more important to understand the marketplace that is being served. Supplying a part that will be incorporated into a larger system differs from selling the whole system. In higher-value cases, many more requirements must be considered. In this chapter, tools are discussed that can be used to come to grips with defining such customer requirements.

One of these tools is Failure Modes and Effects Analysis (FMEA). This tool has excellent applications in defining customer requirements. Another important tool discussed in this chapter is Benchmarking, the subject of many books today. Here, focus is primarily on how Benchmarking can be used to define customer requirements. Before discussing these tools, it is necessary to understand just what information should be defined (see Figure 2.1).

Specified and Unspecified Requirements

✓ How can the customer cause misuse?

✓ What are the life hazard conditions?

✓ What is the service life?

✓ What are the reliability goals?

✓ What are the costs associated with the goal?

✓ What is the customer's use plan?

2.2 Specified and Unspecified Requirements

Approach your customer's specified and unspecified requirements by starting with a life-use plan. How does your customer (whether a direct customer or the customer's customer) plan to use a device? In understanding the life-use plan for the device, some major design considerations can be established.

By looking at the life-use application, an extensive array of specifications can be defined. A simple example can show just how many of your customer's requirements can be found when considering the customer's life-use plan.

Consider the requirements for an IC within a beeper. What are the life-use plans for a beeper and for the IC? First, the Original Equipment Manufacturer (OEM) is going to incorporate the IC into the beeper to process signals or data. Once the IC is incorporated into the beeper, the OEM will functionally evaluate the beeper to ensure that it meets performance requirements. The OEM will most likely perform accelerated qualification tests for the OEM's customers to ensure that manufacturing defects do not slip through the manufacturing process. Once qualification testing has verified that the beeper meets its planned performance requirements, the device will be packed and shipped to a retailer for sale to the end-customer. The user attaches the beeper to a belt to receive messages within numerous types of building structures, and he/she will press a series of buttons to display, save, and delete this information. The user will also subject the beeper to numerous life hazard conditions, such as exposure to the environment in a car, outside on a porch, near a coastal beach, in a polluted environment, and so forth.

Common misuse events that may occur during the life of a beeper include being dropped onto a floor, off a shelf, into a wet sink, being left out in the rain, or occasionally being tossed onto a chair. Service life needs may occur, such as changing the battery. Finally, the customer may specify a certain level of reliability over a period of years. Often, the manufacturer uses Reliability Predictive Modeling (RPM) to assess whether the specified reliability goals are obtainable (see Chapter 11). Table 2.1 summarizes this information.

Table 2.1
Specified and unspecified requirements for a message beeper

Requirement	Useful Tool	Expectation
Customer Use Plan	Benchmarking/ FMEA	• Receives information in numerous building-type structures and environments • Audible beep • Tolerates mechanical wear and tear on beeper buttons and belt clip • State-of-the-art capability
Life Hazard Conditions	FMEA	• Survives numerous polluted environments • Withstands different temperature extremes • Withstands corrosive coastal environments
Common Misuse	FMEA	• Survives numerous shocks from being dropped from heights of up to 6 feet • Survives numerous vibration exposures from normal use and being tossed around • Survives exposure to water (being left out in the rain or dropped momentarily into a wet area, such as a sink or puddle)
Service Life Needs	Benchmarking/ FMEA	• Easy access for battery changes • Mechanical durability of battery clip
Reliability Goals	RPM	• 99% reliability for five-year life

2.3 Cost of Reliability

Almost all of the requirements described in Table 2.1 are directly related to reliability. Meeting and exceeding customer requirements are part of the cost of reliability. Once the requirements have been understood, the challenge is to provide a reliable product that meets these requirements and maintains a competitive product price. Reliability costs are of two types: internal costs, such as reliability assurance and material/design costs, and external costs, such as warranty costs.

External warranty costs imply repair and replacement issues. This can depend on volume. If the volume is extremely high, such as with beepers, repair and replacement issues can become a logistical problem. For example, in Table 2.1, a 99 percent five-year reliability has been specified. This implies that at the end of five years, about 1 percent of the population may fail. If the product has a one-year warranty and the failure rate is constant, about one-fifth of this 1 percent will have failed in the first year. This implies that, for every 1 million beepers shipped, about 2,000 returns could occur each year. Although a return of 2,000 units is not high, the loss of one customer could be expensive. To remain profitable, such costs must be weighed against the initial cost.

Internal costs of ensuring reliability include up-front costs such as FMEA, Process Reliability studies, Design Assessment Reliability Testing, Design Maturity Testing, and screening and monitoring. Other internal costs normally associated with a product are total manufacturing cost of component (higher-rated part) and raw material selections that add to a higher level of reliability but at a higher product cost.

2.4 Benchmarking

Benchmarking is a proactive process for making organizational improvements. In Benchmarking, comparisons are made between a manufacturing process, product, and/or service and industrial best-in-class standards. This process should be ongoing to provide timely comparison of key best-in-class success factors of competitor(s) and to benchmark these factors. Once standards are benchmarked, this information is used to close gaps through identified corrective actions in an organization, process, product, and/or service. Benchmarking has been more commonly used as a key business driver for improving business systems, not as a tool for defining customers' requirements. The benefit of the Benchmarking process is that it helps to focus attention and resources on meeting customer expectations through the continual assessment of industrial benchmarked standards set in the free marketplace. Meeting and/or exceeding these standards ensures customer satisfaction.

2.4.1 Benchmarking to Improve a Business Process

In a simple hypothetical example, a company's high customer satisfaction in the area of telephone orders may be related simply to operator response rather than to product difference. In Benchmarking, analysis results show that this particular business success is correlated to calls that are answered in two or fewer rings. This success factor has a direct impact on customer satisfaction. In this way, Benchmarking can be used to help understand customer expectations and needs, and to set strategies with selected target goals, all without spending valuable time in lessons learned. In this way, Benchmarking can be a key factor in satisfying a customer.

2.4.2 Technology Benchmarking to Improve a Product

Benchmarking is most commonly used to help improve a product. In the Benchmarking process, the market trend, the current best-in-class technology, and the marketable technology for the next generation must be understood. This of course is essential for understanding customers' needs. It allows the voice of the customer to be heard during development of products and services.

For example, in the fast-paced evolution of the computer marketplace, no corporation can operate in isolation with a microscopic picture of the market and expect to be profitable. It would be difficult to be competitive without making continual Benchmarking assessments of computer chip speed, size, cost, reliability, task capability, manufacturability, and so forth. Each step in the evolution of the computer redefines the market and sets the standard based on the customer's expectations. Benchmarking can provide timely information on where and how fast the technology is evolving. This is an important first step in defining customer requirements. From that information, a product's needs may be planned in order to meet the current product requirements. Understanding a product's deficiencies can provide opportunities and rewards in funding research and development efforts. Benchmarking supports customer satisfaction goals by providing timely, cost-effective, high-quality products and service information.

2.4.3 Reverse Engineering to Improve a Product

Many people confuse Reverse Engineering with Benchmarking. Reverse Engineering is only a fraction of the activity involved in performing product Benchmarking. Reverse Engineering is often defined as an improvement tool that helps to assess product reliability performance. This is done by evaluating a competitor's product, carefully disassembling it, and comparing it with the design of interest at each level of the disassembly process. This allows many improvement opportunities for the organization to close the competitive gaps and to expand the company's competitive advantage in the marketplace. In Reverse Engineering, reliability assessment is made by the following: how the competitor's product was used in terms of long service life application; how the competitor's product was built to perform better in terms of device reliability; the number of parts utilized in the product design to achieve target Mean Time Between Failure; the cutting-edge manufacturing approaches utilized in product assembly process; ease of repair; and materials used to obtain product quality/reliability and cost requirements.

This evaluation process is a major step in a company's core effort in assessing product reliability by translating evaluation results into new product features through competitive designs to keep up with today's increasing marketing demands. Reverse Engineering involves extensive competitive product investigation to gain insight into how competitive products are constructed for superior reliability performance. The primary function of Reverse Engineering is to examine products from the best-in-class industrial performers to understand how these competitive products achieve higher success by comparison to your product. The results of the study enable the company to concentrate on corrective actions in the areas where design improvements are needed. *(Note: Caution must be taken not to violate any patent laws when incorporating into your own design what has been learned during Reverse Engineering.)*

2.5 Using Failure Modes and Effects Analysis to Meet Customer Requirements

Traditional FMEA is implemented as a post-engineering activity to check a system. Moving the FMEA process to the Idea Phase allows engineers to optimize numerous design aspects including the Design for Reliability (DfR) process. A product functional-level FMEA can then incorporate the understanding of the customer's needs with product capabilities. This FMEA requires the design team to walk through the design and define what the product's functions should be and how the product should function. Many engineers in the design community have been doing this for years. However, using FMEA gives the design team a formal process to collect information concerning a customer's needs while ranking the importance of each design requirement. This ranking system allows the design team to make a tradeoff between the design's capability and the customer's needs. Such simple preparation has become more and more critical in industry with today's time-to-market pressures and the need for a successful, streamlined technological design that provides customer satisfaction.

Customers have concerns that are typically overlooked. For example, it is common to overlook Mean Time Between Failure (MTBF) and maintainability (maintenance concept for field support) goals. A specific concern is that the product maintenance characteristics enable effective user fault detection and fault isolation. The feasibility of meeting the desired MTBF in order to establish internal reliability program tasks (i.e., DfR) that ensure a product will meet the desired characteristics must be estimated. This is also the point in early program development to identify if the reliability growth should be highly aggressive. This can often be met with an accelerated reliability growth stage gate program. Another overlooked customer concern is the user's application, necessary system functions, and the intended use conditions. These are all critical. Specifically identifying both the use and the possible misuse conditions is a concern. It may be decided to perform an FMEA to evaluate this characteristic more formally. If this is in a safety-critical application, has an approach to control the safety been established? It may be decided to perform a Customer Safety/Hazard Analysis to evaluate this characteristic more formally.

Understanding a modern customer's full requirements can be very complex. As higher-value solutions are provided, it becomes important to understand the marketplace that is being served. Supplying a part that will be incorporated into a larger system differs from selling the whole system. In higher-value cases, many more requirements must be considered.

CHAPTER 3

Design Assessment Reliability Testing

3.1 Introduction

In Design Assessment Reliability Testing (DART), product risk-mitigation studies are performed using primarily highly accelerated test methods. The objective of these studies is to mature the design as fast as possible through early identification of potential failure modes and corrective-action measures (see Figure 3.1). This phase is essential as a primary tool in the highly accelerated growth process.

The reliability growth of most products occurs in this phase (see Figure 3.2). Here, the cost of reliability growth fixes is much less than in the next Design Maturity Testing (DMT) Phase (see Chapter 4). Therefore, it is essential for products to mature in this phase.

In the Idea Phase (Stage gate 1), where an idea for a product is just solidifying, no real growth can occur. The only major impact on product reliability is selection of the platform on which the product is built.

The Evaluation Phase (Stage gate 2) is prior to having all major design components selected and finalized. For potential major design modifications, the Evaluation Phase will have the most effect on a product's reliability.

In the Development Phase (Stage gate 3), Transition Phase (Stage gate 4), and Production Phase (Stage gate 5), hand-tooling is in place, circuit-board layout is completed, and component selection has been finished.

The goal of Design Assessment Reliability Testing is to identify any potential failure modes that are inherent in a design early in the design process. By identifying the root cause of the failure mode and then incorporating a fix to the design, reliability growth can be achieved. This is accomplished by designing out the possibility of potential failure modes occurring with the customer and reducing the inherent risk associated with new product development. This process is also known as the Risk-Mitigation Phase. Figure 3.1 demonstrates the process and some of the added benefits of the Design Assessment Reliability Testing.

Design Assessment Reliability Testing at the assembly or subassembly level utilizes step-stress testing as its primary test method. It should be noted that Highly Accelerated Life Testing (HALT) is not meant to be a simulation of the real world but a rapid way to stimulate failure modes. The mathematics and concepts of step-stress testing are described in Chapter 9. At the hybrid or component level, alternate test methods may apply as described in Section 3.3. These highly accelerated programs can find potential device failure modes in the fastest possible time. These methods commonly employ sequential testing, such as step-stressing the units with temperature and then vibration. These two stresses can be combined so that temperature and vibration are applied simultaneously. This speeds up testing, and if an interactive vibration/temperature failure mode is present, this combined testing may be the only way to find it. These methods are discussed

Figure 3.1
Design Assessment Reliability Testing

Figure 3.2
Accelerated reliability growth program

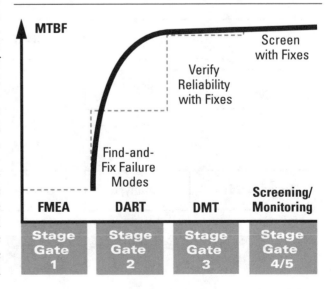

below. Other stresses used may be power step-stress, power cycling, package preconditioning with infrared (IR) reflow, electrostatic-discharge (ESD) simulation, and so forth. The choice depends on the intended type of unit under test and the unit's potential failure modes.

3.2 Four-Corner HALT Testing

HALT is primarily for assemblies and subassemblies. The HALT test method utilizes a HALT chamber. Today, these multistress environmental systems are produced by a large number of suppliers. The chamber is unique and can perform both temperature and vibration step-stress testing. This testing is primarily performed in the Evaluation Phase where early breadboard units and prototypes are built (see Figure 3.3). Testing is also performed in stage gate 3 to help verify previous HALT test results.

Throughout the test phases, Failure Analysis (FA) and Corrective Actions are performed. Since HALT is highly accelerated, reliability growth occurs rapidly once corrective action fixes are properly incorporated. Additionally, such testing helps establish potential screening criteria that may be necessary in the first year of product maturation.

In the HALT test method, four-corner testing is commonly done. First, temperature and vibration step-stress tests are performed, as described in Sections 3.2.1 and 3.2.2, respectively. The purpose of these tests is to find both the operating and nonoperating failure limits of the unit(s) under test. Such limits, when found for vibration and temperature, establish the four corners of the HALT test (see Figure 3.4).

When failure modes are identified, engineers assess the failure modes and perform appropriate corrective action. Common-sense guidelines should be followed in assessing failure modes as outlined in Section 9.2. Additional testing is conducted that combines temperature and vibration, as described in Section 3.2.3. Finally, a rapid thermal stress test is performed as described in 3.2.4. After corrective actions have been applied, units are tested in a similar manner to determine the effectiveness of a fix, to identify any new failure

Figure 3.3
Stage gate overview with HALT

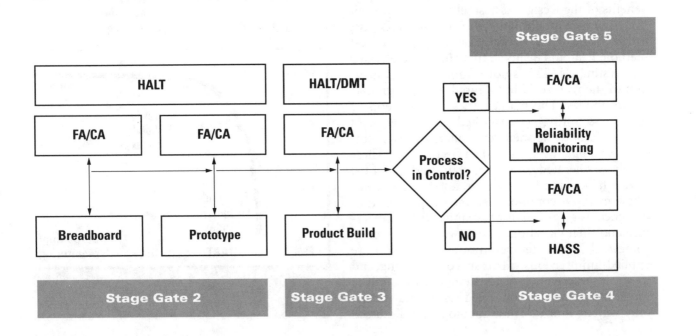

mode, and to establish new limits. Any realistic associated failure modes that are identified must again be studied through the failure analysis/corrective action system.

Figure 3.4
HALT testing

3.2.1 HALT Thermal Step-Stress

Probably the most common step-stress test is temperature. HALT utilizes temperature step-stress testing to activate temperature-dependent failure modes. In general, step-stress is an alternative to life testing (see Chapter 9). It is a very powerful tool for finding nonmoisture thermochemical mechanisms (e.g., metal interdiffusion, intermetallic growth problems such as Kirkendall voiding, electromigration, MOS gate wearout, etc.). The test is also used for in-depth reliability studies (see Chapter 9). As a HALT tool to probe for potential failure modes, thermal step-stress tests start at room temperature and usually apply cold step-stresses by decreasing the temperature nominally in 10°C steps. Each ther-

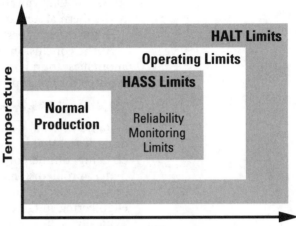

mal dwell is long enough for the unit to stabilize in temperature. In HALT, units are monitored so that the lower operational limit can be pursued. Fixes take place on all failure modes. Testing continues until it is determined that the observed failures are due to wearout mechanisms.

Then the hot step-stress is performed with increasing temperature nominally in 10°C steps starting from room temperature. Each thermal dwell is long enough for the unit to stabilize in temperature. The unit's upper operational limit is pursued with repairs taking place at all relevant failures until it is determined that the failures are due to wearout mechanisms.

Once the operational limits are identified, the destructive limits are found: first the cold limit, then the hot limit. The operational limit is used to identify the point at which the unit goes out of operational specifications but can operate with a slight temperature adjustment. The destructive limit is used to identify the point at which the unit will not recover, even with the removal of all stresses.

3.2.2 HALT Vibration Step-Stress

Application of vibration stress is an important tool for finding mechanical failure mechanisms such as problems related to mechanical attachment, package integrity, fatigue, etc. Vibration step-stress testing is run by increasing the vibration level from zero in predetermined steps (based on engineering judgment) over a specified frequency region. The input vibration is continually stepped until operational and/or destructive levels for the unit are obtained. The dwell time of each vibration is predetermined prior to testing. Tickle vibrations of a predetermined level may be applied for the detection of failures precipitated but not detected during higher vibration levels.

3.2.3 Thermal/Vibration Step-Stress

The rationale behind combining stresses in HALT is the belief that interactive temperature/vibration failure mechanisms exist. Such mechanisms may not be easily detected with the application of only one stress. Potential failure mechanisms, when exposed to both environmental stresses simultaneously, are subject to a higher degree of acceleration. This applies to mechanisms, such as a number of fatigue failure modes, which are influenced by both environments (temperature and vibration) independently. However, in many cases, one stress or another can detect failure modes without the need for a combined test.

In the combined HALT test, the top six temperatures of the 10°C thermal step-stress test are run. The combined vibration level starts at approximately half of the "operational" vibration level discovered during the vibration step-stress test. One vibration level is applied for each of the six lowest temperatures of the thermal step-stress test. The rapid thermal transition test will then be run with the three highest vibration levels discovered in the vibration stress test.

At the conclusion of the test, a failure analysis report is provided detailing anomalies found during the HALT process. Corrective actions are then assessed for the appropriate fixes that are needed to provide a robust design.

3.2.4 Rapid Thermal Transitions

Rapid thermal transition is not a redundant test. It is a temperature shock or cycle test that is used to find thermomechanical mechanisms (e.g., package cracking, ohmic contacts, wire bond/lead integrity, thermal expansion mismatch problems, metal fatigue, etc.). Thermal transitions are performed between selected low and high temperatures that are determined based on engineering judgment and results from the thermal step-stress test analysis. Typically, values are decreased from operational limits. Generally, thermal transition rates are performed as fast as the chamber and product/mass thermal stabilization will allow. Therefore, thermal dwells are held long enough for the chassis temperature to reach the thermal setpoint. The unit is powered during transitions from cold to hot temperatures and at the dwell temperature, and off during the hot-to-cold transition. Four thermal cycles (eight transitions) are generally performed.

3.3 Design Assessment Reliability Testing at the Hybrid and Component Level

Figure 3.5
Design Assessment Reliability Testing at the hybrid and component level

Design Assessment Reliability Testing can also be performed at the hybrid and component level. This testing may or may not involve step-stress testing. It must be specifically designed for the platform of interest.

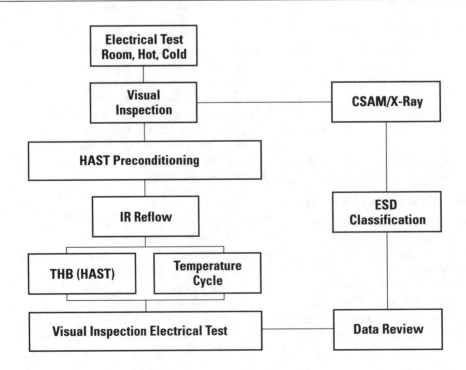

Figure 3.5 illustrates a program that was set up for a plastic-encapsulated hybrid. Essential in any Design Assessment Reliability Testing program is the time frame and quality of testing to take place. For example, the program in Fig. 3.5 can be completed in less than two weeks. Of course, actual cycle time depends on testing availability. The tests described here provide a level of engineering confidence that units will pass Design Maturity Testing, which normally follows the Design Assessment Reliability Testing stage gate. For example, testing here includes a check on preconditioning, IR reflow, temperature cycling, ESD, temperature-humidity-bias tests, and so forth.

3.4 Summary

A Design Assessment Reliability Testing program is an essential ingredient for any successful reliability program in today's semiconductor world. Many industries rely solely on such a program, using it as their only phase. However, in a stage gate approach, Design Assessment Reliability Testing is only one gate and should not be used to substitute for the entire reliability growth program. The full reliability program is necessary for meeting and exceeding customer expectations.

CHAPTER 4

Design Maturity Testing (DMT)

4.1 Introduction

This chapter discusses the Development Phase through the application of reliability Design Maturity Testing (DMT) for ICs, hybrids, and assemblies to ensure high reliability and long-life (10 years) applications. These components and assemblies represent a major portion of products in the IC industry (see Figure 4.1).

DMT should be based on statistically driven sampling plans, primarily for environmental testing, to ensure that products are robust against life hazard conditions that could cause catastrophic field failures. Verification includes accelerated testing, which must be performed to save time and money while assuring product reliability. Since major decisions are based on such tests, planning must ensure statistically significant testing at the desired level of confidence.

This chapter describes DMT methods as well as the associated models for accelerated testing, statistical confidence, and reliability success-and-failure probability confidence bounds. Test procedures maintain state-of-the-art testing needs associated with potentially new products, best commercial practices, new equipment capabilities, and new understanding of potential failure mechanisms.

The main objective of DMT is to determine whether the design will meet its reliability objectives. DMT is planned to:

- Demonstrate a level of a product's reliability;
- Provide statistically significant test plans that reasonably balance reliability protection and add-on costs; and
- Provide a plan that gives guidance to engineers for verification tests and acceptance criteria.

Planning documentation for DMT should include the basic reliability physics behind DMT in order to provide engineering flexibility in verification analysis.

4.2 Overview of DMT Planning

DMT demonstrates a certain level of product reliability to help verify a product's failure rate. Reliability objectives help to specify design verification requirements and meet customer requirements.

Four main accelerated verification tests are performed: High-Temperature Operating Life (HTOL), Temperature Cycle, Vibration, and Temperature-Humidity-Bias (THB) (see Figure 4.2). These tests and their objectives are discussed in more detail below. Each test stresses to some degree most failure modes. However, each test is historically known to cause higher stress levels for certain failure modes. Therefore, reliability must be allocated properly for each test. This means that when a quantitative reliability objective is established for the product, it is further subdivided. For example, in certain technology THB, Temperature Cycling, and HTOL failure modes have been found to compare to approximately 20 percent, 30 percent, and

High-Temperature Operating Life	Temperature Cycle
Intermetallics, Diffusion, Elec. Overstress...	Thermal-Mechanical Connections, Materials...
Temperature-Humidity-Bias	**Vibration**
Process-Related Corrosion, Dendrite Growth, Leakage Current...	Structure Durability, Resonances, SMT...

50 percent, respectively, of the total reliability, while some tests such as Vibration may be designed to demonstrate capability at the system level. Reliability objectives, specific test requirements, and guidelines depend on the technology platform on which the design has been built. Therefore, the development of proper DMT planning requires a unique understanding of the technology platform.

4.3 DMT Reliability Objectives

DMT documents specify certain reliability objectives (see Figure 4.3). Objectives 1, 2, 3, and 4 may be sequentially less extensive in terms of reliability requirements and test specifications. This does not necessarily mean, however, sequentially less reliability, since the long-term reliability of each category and test will generally be much better than the stated objectives. Test requirements provide high-reliability assurance since DMT is usually specified in accordance with the zero-failure test procedure described below. Guidelines for choosing the appropriate objective depend on customer expectations and on product capability. Generally, all things being equal, as the number of components for hybrids and assemblies increases, so, too, does the intrinsic failure rate. Thus, the objectives can have guidelines based on part counts for hybrids and assemblies. However, a reliability predictive model, such as military standard (MIL STD) 217 or Bellcore analysis, is recommended in making the actual assessment (see Chapter 11). Additionally, DMT usually requires that all products pass Objectives 1, 2, 3, or 4 via a zero-failure test procedure.

4.4 DMT Methods

Product maturation should include both process and product reliability. Prior to full-scale manufacturing of any new products, beta products should pass Design Assessment Reliability Testing similar to that in Chapter 3. This phase includes Highly Accelerated Life Testing (HALT). By the time a product is ready for DMT, the process should be fairly mature. After passing the Design Assessment Reliability Testing stage gate, DMT is performed. Assemblies are a good example of the DMT process since they are complex products. As Figure 4.4 shows, assembly testing typically consists of 15 blocks (B1–B15). Each document has a process card. In order to implement testing on a practical scale for the vast number of device projects, software programs can be provided to plan testing for plastic-packaged ICs, discretes, hybrids, and assemblies. More information on these programs is provided below. Once automated, the process card follows the blocks in Figure 4.4.

Figure 4.3
Typical DMT objectives

In Block 1, initial samples are provided for performing DMT (sampling is described below). Then visual and electrical go/no-go tests are performed in

Reliability Objective	Plastic ICs/ Discretes (FITs)	Hybrids (FITs)	Assemblies (FITs)
1	5	400	400
2	50	1,000	1,000
3	100	2,000	4,000
4	400	4,000	10,000

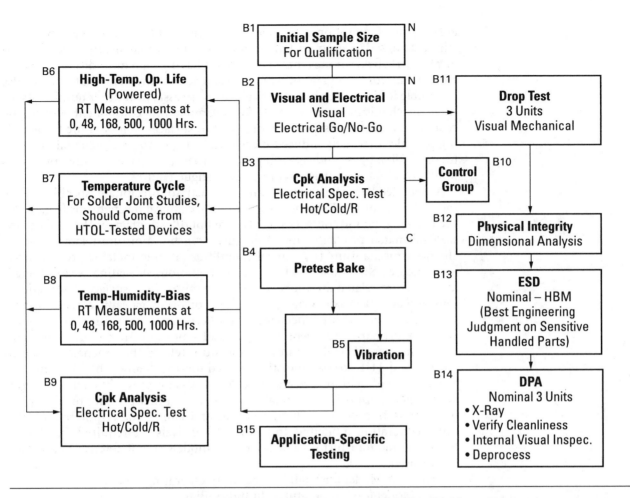

Figure 4.4
Typical DMT for SMT/PC-based assemblies

Block 2. This step can be combined with Block 3 for process capability (Cpk) analysis. The purpose of process capability analysis in Blocks 3 and 9 is to verify that all key electrical parameters remain within the process capability indices limits over both their specified temperature ranges and over accelerated-life-test conditions. This test is performed at both upper and lower specified temperature conditions and at room temperature. Statistical analysis ensures the distribution behaves normally and within reasonable Cpk limits. DMT documents should call for a Cpk test before and after accelerated life testing. Accelerated life tests nominally look for catastrophic failures; however, parametric degradation can also be a source of device failure.

A Cpk test performed before and after accelerated testing helps engineers assess whether significant parametric change has occurred as a result of aging. If so, corrective actions such as an appropriate burn-in period may be needed to help remove parametric aging problems. After the first Cpk test, samples can follow two paths: accelerated testing in Blocks B4 to B8 and nonaccelerated testing in Blocks B10 to B15.

4.4.1 Summary Description of Accelerated Tests (Blocks B4 to B8)

Blocks B4 to B8 provide an overview of basic verification testing that includes accelerated temperature control, THB, HTOL, and Vibration testing (see Figure 4.4). Historically, the failure mechanisms influencing the long-term reliability of assemblies have been processes that are relatively strong functions of temperature, humidity, operating voltage (or current), and mechanical vibration.

The HTOL test having high temperature generally promotes diffusion mechanisms, metal migration, and annealing processes. Electrical bias is required to stimulate a wide variety of temperature-sensitive mechanisms dependent on local electric fields or current. The THB test having high humidity can accelerate galvanic corrosion and other chemical reactions involving water molecules or soluble ionic species, but electrical bias is necessary for other humidity-sensitive mechanisms such as material delamination, component electrolytic corrosion, and charge separation of insulator surfaces. Temperature Cycle testing provides thermally induced mechanical stresses to precipitate failures due to material fatigue and assembly faults. Lastly, vibration testing promotes nonthermal-induced stress to precipitate fatigue failures.

Accelerated verification tests are designed to precipitate specific failure modes/mechanisms. Examples are thermomechanical mechanisms (e.g., package cracking, ohmic contacts such as wire bond/lead integrity, thermal expansion mismatch problems, metal fatigue, etc.), nonmoisture-related thermochemical mechanisms (e.g., metal interdiffusion, intermetallic growth problems such as Kirkendall voiding, electromigration, sidegating wearout, etc.), and moisture-related thermochemical mechanisms (e.g., surface charge effects, ionic leakage effects, dendrite growth, lead corrosion, galvanic corrosion, etc.).

To stress each failure mechanism properly, all four accelerated tests are required. Primarily, Temperature Cycle and Vibration stress thermomechanical mechanisms, HTOL stresses nonmoisture-related thermochemical mechanisms, and THB stresses moisture-related thermochemical mechanisms.

The concept behind accelerated testing is to compress time and accelerate failure mechanisms in a reasonable test period so that product reliability can be assessed. In order to evaluate test time and sample size requirements, both time acceleration modeling and statistical analysis are required. For typical test planning, four historical acceleration models (see Reference 1) are commonly used:
- Arrhenius Model for High-Temperature Operating Life,
- Peck Model for Temperature-Humidity-Bias,
- Coffin-Manson Model for Temperature-Cycle, and
- Power Spectral Density Power Law for Vibration.

These are described further in Chapter 9.

4.4.2 Nonaccelerated Testing (Blocks B10 to B15)

Blocks B10 to B15 (see Figure 4.4) provide an overview of design maturity nonaccelerated tests. A control unit is necessary to validate measurement accuracy over DMT in Block B10. The control unit does not undergo any test other than electrical measurement. For assemblies, a drop/shock test is performed in Block B11 to assure that material can withstand the relatively infrequent, nonrepetitive shocks or transient vibrations encountered in handling, transportation, and service environments. Units are also required to meet all

Table 4.1
ESD classification

ESD Class	Applied Voltage	Requirements*
Class 0	200 V	One or more units fail
Class I	1,000 V	Pass Class 0, one or more units fail from 201 to 1,000 V
Class II	2,000 V	Pass Class I, one or more units fail from 1,001 to 2,000 V

Three positive and three negative pulses applied to all external leads

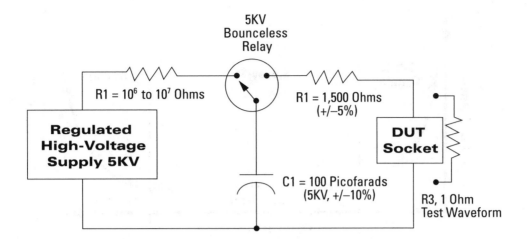

Figure 4.5
*ESD classification
test circuit
(Human Body Model)*

dimensional specifications in Block B12. DMT should include physical integrity to ensure that the design is mature and that simple problems such as dimensionality will not cause customers unforeseen problems.

Electrostatic Discharge (ESD) testing is performed in Block B13 to determine the susceptibility of an assembly and its associated components to ESD sensitivity. In addition, units may be classified as to their sensitivity (see Table 4.1). It is common to classify assembly sensitivity using a Human Body Model ESD test (see Figure 4.5) that applies ESD pulses to exposed external leads. Other common ESD tests that can be used are Charge Device Model or Machine Device Model ESD testing.

DMT also includes the destructive physical analysis in Block B14. Units undergo internal visual inspection and a cleanliness check to ensure that workmanship is adequate and that contaminants or unwanted assembly defects are not present. An x-ray study will ensure that packaged units and molded parts appear to be constructed properly. Finally, units will be subject to deprocessing to review how well the assembly is put together. Any application-specific tests should also be performed in Block B15. Application-specific tests can include flammability, Federal Communications Commission (FCC) compliance tests, FCC Electromagnetic Interference (EMI) tests, and so forth.

4.5 Reliability and Sampling Distribution Models

Life testing often furnishes few failures. Historically, it has been customary to model a failure population with a log-normal or Weibull distribution for components. However, when anticipated life test results are expected to produce few catastrophic events, sampling plans are often, as in the case here, based on zero-failure (or failure-free) testing.

In this case, it is reasonable to model reliability using the exponential distribution for two reasons: failure distributions cannot be determined, and reaching wearout is not anticipated on high-reliability ten-year verification testing. This indicates that verification aging will still be in the steady-state portion of the reliability life model, known as the bathtub curve (see Figure 4.6). The steady state is modeled using an exponential distribution where the failure rate, λ, is constant and the reliability function, R, over time, t, is

$$R(t) = e^{-\lambda t} \qquad (4.1)$$

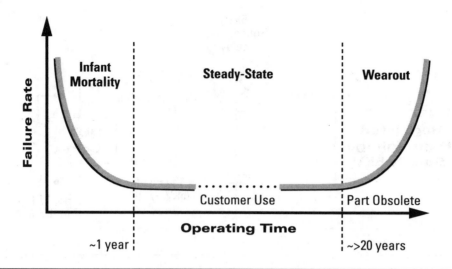

Figure 4.6
Reliability bathtub curve model

The fraction of devices that fail at or before time t is $F(t) = 1 - R(t)$. Chapter 8 provides more details on this distribution. An additional reason for using the exponential distribution is that a constant failure rate model is usually used for complex systems/assemblies. An exponential failure rate depicts random failure events that occur in complex systems.

4.6 Sample Size Planning

Once reliability objectives are established (see Figure 4.3), sample sizes can be planned with statistical confidence. To determine a statistical sample size to meet a particular reliability objective at a specific confidence level, it is common to use a chi-square, χ^2, confidence estimate, given by

$$N = \frac{\chi^2(\gamma,\ 2Y+2)}{2\bar{\lambda}At} \qquad (4.2)$$

where
  $\bar{N}$ = the sample size,
  λ = the upper-bound or failure-rate objective,
  Y = the number of failures (nominally zero),
  t = the total test time,
  A = the estimated test/failure mode acceleration factor, and
  γ = the confidence level (nominally 90%).

In the chi-square estimator, the failure rate objective is taken as the single-sided upper-bound estimate. To demonstrate a reliability objective, a number of factors are required, including the total test time, an estimate for the environmental acceleration factor, the confidence level, and the number of allowed test failures. If the number of allowed failures is zero, it is termed "zero" or "failure-free" testing. Since the sample size depends on an estimate for the acceleration factor, conservative estimates should be used based on historical information and process reliability studies. A failure-free accelerated test plan example is provided in Chapter 9, Section 9.10.

As an example, if DMT Objective 1 is met for assemblies at the 90 percent confidence level, we can be 90 percent confident that the failure rate is no higher than

1,000 FITs and that the one-year reliability (lower bound) is no lower than

$$\underline{R(t)} = e^{-\overline{\lambda}\,t} = 0.991 \qquad (4.3)$$

Further mathematical details are provided in Chapter 8. In rare instances, such as predevelopment (beta) devices, other sampling schemes may be of interest and are often based on customer requirements for beta devices.

4.7 Automated Accelerated Test Planning

Numerous products/platforms must pass through the DMT as described earlier. To make DMT planning feasible and maintain consistent test methods on all products, it is advisable to use automated test planning software. Historical acceleration models, conservative estimates of model parameters, and statistical sample planning requirements can be included in a fully automated plan, which can also address issues of test duration, test strengths, and customer usage. Automated software also helps to meet ISO-9001 requirements for consistent corporate planning. Such software programs are needed to provide test plans for all device families, plastic ICs and discretes, hybrids, and assemblies. Once automated, engineers can simply input environmental specified requirements and answer test questions (see Figure 4.7). Once input requirements, such as test duration, test conditions, customer-specified conditions, and the specified failure rate objective are made, software can provide a statistically significant DMT with an output process card. The process card is a step-by-step test procedure required to perform the DMT. The software automatically takes the inputs and estimates the optimal statistically significant sample size required to meet the input failure rate objective at a specified confidence level. The process card should also indicate the test conditions, test times, and guidelines for completing each test specified on the card.

Figure 4.7
Automating DMT

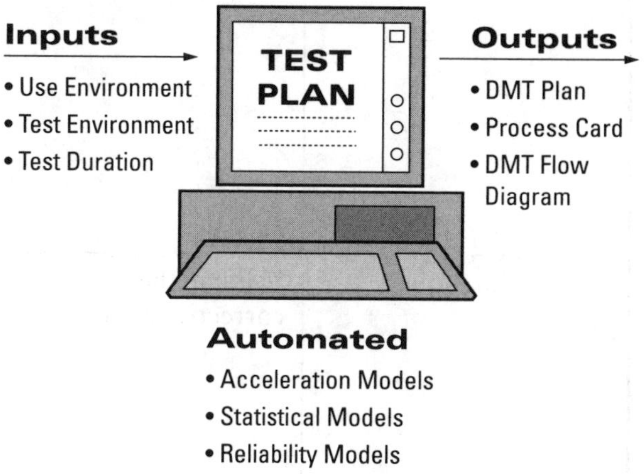

Inputs
• Use Environment
• Test Environment
• Test Duration

TEST PLAN

Outputs
• DMT Plan
• Process Card
• DMT Flow Diagram

Automated
• Acceleration Models
• Statistical Models
• Reliability Models

4.8 DMT Methodology and Guidelines

DMT is described at specific conditions for stress levels, time duration, and the number of units that must pass the test in order to meet a specific reliability failure rate objective (e.g., 1,000 FITs). If all units pass the failure-free test period at a 90 percent confidence level, for example, we will be 90 percent confident that the product's failure rate is no higher than the specified failure rate objective of the test. In this section, we describe some guidelines and methodology for DMT.

4.8.1 Guidelines for Zero-Failure Testing

To provide statistical confidence in reliability, the most efficient planning is based on failure-free testing. Statistical confidence levels are used to make inferences about a population, given data from a sample. It describes the fraction of the times the confidence level will capture the true value in repeated tests. A 90 percent confidence level is common in best commercial practice. Failure-free

Procedure	Description	Guidelines
Root-Cause Failure Analysis	Identify and analyze the modes, mechanisms, causes, and consequences of potential and real failures	All failures should undergo a root-cause analysis. This may include failure analysis. Once identified, the failure mode shall be classified as Type A or Type B. Type A modes are not fixable and no corrective action can be taken because it is not cost-effective. In this event, contact corporate reliability. Type B modes are fixable failure modes by design, process, or workmanship change.
Corrective Action	Correct the root cause of the failure and implement the change	Determine the best possible fix in the shortest amount of time. All corrective actions should be evaluated prior to implementation. This should include a peer review by a concurrent engineer design team including one or more reliability engineers. An estimate should be made of the effect of the proposed fix on reliability (MTTF), and the fix should be evaluated on a scale from 0 to 1.0, where 1.0 implies that the fix will eliminate the failure mode. If the fix has a low rating (<0.6), other possible fixes shall be considered and rated until the best possible solution is found.
Validation	Validate the corrective action	If time and money permit, the design should be fully retested with a statistically meaningful sample size as described in this document. If time and money do not permit, units should undergo partial testing to demonstrate capability with the approval of reliability engineering. If necessary, units may still require the implementation of a Test-Analyze-And-Fix (TAAF) reliability growth program. Early production units should enter the reliability growth screening program.

Table 4.2
Procedures for zero-failure testing

testing is usually planned because it is the most efficient test in terms of saving time and money while demonstrating reliability. However, we know that in the real world, things can go wrong. *What happens if one or more units fail?* Common guidelines proper for the action of implementing appropriate fixes with reliability growth are presented in Chapter 10, which fully discusses traditional Crow-AMSAA reliability growth (see References 2 and 3) test methods. These methods consist of failure analysis, corrective action, and validation.

Such testing permits quantification of the increase in reliability due to fixes. This results in improved product reliability over time as a result of the iterative process of testing and identification and correction of design flaws, part defects, and/or workmanship defects. The best commercial reliability growth practice for a unit failing zero-failure testing is to follow three simple steps: root-cause failure analysis, corrective action, and validation. See Table 4.2 for the details of these three steps. In some instances, if fixes do not immediately solve the problem, a production-screening program may be necessary. Once corrective actions are implemented, production monitoring may be necessary to ensure that fixes are properly in place. In the next chapter, we describe the Production Screening and Monitoring stage gate.

References

1. Nelson, Wayne, *Accelerated Testing: Statistical Models, Test Plans, and Data Analysis,* John Wiley & Sons, New York, 1990.

2. Department of Defense, "Military Handbook-189, Reliability Growth Management," *Naval Publications and Forms,* Philadelphia, 1981.

3. Feinberg, A. A., and Gibson, G. J., "Accelerated Reliability Growth Methodologies and Models," *Recent Advances in Life-Testing and Reliability,* Edited by N. Balakrishnan, CRC Press, 1995.

CHAPTER 5

Screening and Monitoring

5.1 Introduction

If all processes were under complete control, product screening or monitoring would be unnecessary. If products were perfect, there would be no field returns or infant mortality problems, and customers would be satisfied with product reliability and quality. However, in the real world, unacceptable process and material variations exist. Product flaws need to be anticipated before customers receive final products and use them. This is the primary reason that a good screening and monitoring program is needed to provide high-quality products. Screening and monitoring programs are a major factor in achieving customer satisfaction.

Parts are screened in the early production stage until the process is under control and any material problems have been resolved. Once this occurs, a monitoring program can ensure that the process has not changed and that any deviations have been stabilized. Here, the term "screening" implies 100% product testing while "monitoring" indicates a sample test. Screens are based upon a product's potential failure modes. Screening may be simple, such as on-off cycling of the unit, or it may be more involved, requiring one or more powered *environmental stress screens*. Usually, screens that power up the unit, compared with nonpowered screens, provide the best opportunity to precipitate failure-mode problems. Screens are constantly reviewed and may be modified based on screening yield results. For example, if field returns are low and the screen yields are high (near 100 percent), the screen should be changed to find all the field issues. If yields are high with acceptable part per million (PPM) field returns, then a monitoring program will replace the screen. In general, monitoring is the preferred stage gate for low-cost/high-volume jobs. A major caution that must be given when selecting the correct screening program is to ensure that the process of screening out early life failures does not remove too much of a product's useful life. Manufacturers have noted that, in the attempt to drive out early life failure, the useful life of some products can become reduced. If this occurs, customers will find wear-out failure mechanisms during early field use.

In this chapter, some of the more common approaches to product screening used in industry are discussed.

5.2 Achieving Reliability Growth in a Screening Program

Although most Reliability Growth occurs during the evaluation phase of the design cycle, some benefits can be found when screening and Reliability Growth programs are combined.

Reliability improves when corrective-action fixes are incorporated into products with a screening program. Improvements lower the risk of excessive field return costs. Once failure modes have been effectively eliminated, a monitoring program can replace screening.

Chapter 10 provides more information on estimating Reliability Growth. If fixes are not incorporated during production screening, only the *infant mortality* failures can be removed, and no actual reliability improvements can be made in the steady-state failure rate, which remains constant throughout customer usage. Merely screening a product without incorporating fixes does not increase product Reliability Growth. In Chapter 10, Table 10.1 provides an overview of estimated reliability benefits when incorporating a screening Reliability Growth program.

5.3 Monitoring and Screening Tools

In this section, the main tools in implementing a screening and monitoring program are discussed. For an overview of these tools, see Table 5.1

5.3.1 Thermal Cycling

The most common screens used today are thermal cycling and thermal shock. The thermal cycle test is one in which units are cycled repeatedly between two temperature extremes. Cycling transitions between temperatures are slow (typically 3°C to 5°C per minute) compared with thermal shock. Devices cycled in this way experience expansion and contraction effects, which promote fatigue-related failures. Slow cycling between temperature extremes gives time for other temperature-related effects, such as creep and intermetallic formation, to occur. Primary effects are solder joint creep and interconnection fatigue problems. Because of the slow transitional rates, usually only 5 to 10 thermal cycles are performed in a test. Thermal-cycle testing is performed in a single chamber, and thermal-shock testing is performed in a double chamber. This makes it much easier to measure *in situ* temperature coefficients in a temperature cycle chamber.

Table 5.1

Screening and monitoring tests

Thermal cycling is commonly performed to find problems in surface-mount-technology (SMT) components, ball-grid-array (BGA) solder interconnections, temperature mismatch mounts, contamination under bonds, and hermetic seals. Examples include: temperature mismatch problems in rib-

Screens/Monitoring Tests	Description
Thermal Cycling	Finds fatigue-related defects. Test uses slow temperature transition often to allow for creep-related problems and temperature transitional analysis. Units are easily monitored electrically.
Thermal Shock	Finds fatigue-related defects. Test uses fast temperature transition for expansion/contraction-related defects. Units are not easily monitored electrically.
Stabilization Bake	An unbiased high-temperature bake to remove infant mortality.
Burn-In	A biased high-temperature bake to remove infant mortality.
Fine Leak	Tests fine leaks in hermetic packages.
Gross Leak	Tests gross leaks in hermetic packages.
Particle Impact Noise Detection (PIND)	Primarily used to detect presence of loose particles, such as solder balls, in hermetic packages.
Ionic Contamination	Tests corrosion-related surface contamination of products.
Environmental Stress Screening (ESS)	One or more environmental tests for screening products prior to shipment.
Vibration/Constant Acceleration	Tests mechanical integrity of mounted components.
Combined Stresses	Tests interactive environmental problems.
Humidity	Tests for corrosion-related problems.
Highly Accelerated Stress Test (HAST)	A humidity environmental test to look for failures in the shortest time possible.

bon bonds when bonds lack proper stress relief, solder joint failures when intermetallics are present, surface-mount failures when process variations exist in solder volume, etc. Thermal cycling has become one of the most popular tests, primarily because of its effectiveness as an environmental screen in surveying a wide variety of issues.

5.3.2 Thermal Shock

Thermal shock, similar to thermal cycle, exposes a product to two temperature extremes. Unlike thermal-cycle testing, thermal-shock testing usually uses a double chamber, each set at one of the temperature extremes, as the product is transported between the two chambers.

One chamber may be at –65°C and the second at 125°C. The product is then moved automatically between two chambers that are fitted with a lift-and-seal system, so the product is not exposed to room-temperature conditions. The chamber temperature recovers in about 5 minutes, as measured in a supply air stream. Although the time can vary depending on a product's mass, it typically takes about 15 minutes to reach chamber temperature. Once the product has reached the chamber temperature, it is allowed to dwell for at least 10 additional minutes. This exposes the unit to alternate extreme conditions, promoting expansion and contraction effects within the product. The overall effect is to accelerate environmental effects that occur in the field due to daily or seasonal temperature swings. Thermal shock as a screen has advantages over thermal cycling in that it can automatically cycle large product volumes between temperature extremes quickly and complete each cycle in less than one hour.

Some of the effects encountered as a result of thermal-shock testing are cracking, crazing, delamination of materials and finishes, cracking and crazing of seals and encapsulated compounds, seal leaks in hermetic packages (e.g., metal-to-metal or glass-to-metal seals), and other fatigue-related failure problems. Electrical characterization is usually not performed *in situ* but done before and after testing. Electrical parameter changes in the performance of a product due to thermomechanical problems are checked here. Similar to thermal cycling, this test is most commonly used to accelerate field life conditions. Chapter 9 describes an accelerated testing model that may be used to estimate test time compression.

5.3.3 Stabilization Bake

Stabilization bake is used to determine the unbiased storage effect on microelectronic devices. Such effects can cause electrical performance changes. Often this bake is used as an infant mortality screen. Typically, changes in the electrical performance over test time are monitored to assess unacceptable performance degradation. Intermediate measurements are used to evaluate progress, and such measurements can help refine testing requirements. To perform the test, units are put in an elevated temperature environment for an extended period of time. Chapter 9 describes an accelerated testing model that may be used to estimate the accelerated effects of the test relative to field use conditions.

5.3.4 Burn-In

Many components are burned-in prior to shipping to the customer. The burn-in process consists of high-temperature storage under bias stress. The test is used to screen out borderline units that would otherwise be early-life (first-year) failures. Similar to the stabilization bake, all units are exposed to an elevated temperature to accelerate the failure mechanism. Unlike the stabilization bake, units are biased *in situ*.

5.3.5 Fine and Gross Leak

The fine leak test measures hermeticity effectiveness of components such as hermetic hybrids. This is typically done by placing a device that has been exposed to helium (for a known pressure and time duration) in a mass spectrometer capable of reading how much helium is being emitted by the product over time to estimate the leak rate. The acceptable leak rate is then estimated at normal conditions. The effective range is 10^{-4} to 10^{-10}cc/sec He. This is typically done as a screen (i.e., on 100% of the product) to ensure that products have no hermeticity problems. The gross leak test is for leaks greater than 10^{-4}cc/sec. Some of the methods that are utilized are perfluorocarbon gross leak, penetrant dye gross leak, and weight gain gross leak. A description of the perfluorocarbon and penetrant dye gross leak tests follows. Submerging the device under test to a Type 1 perfluorocarbon fluid in a pressure vessel performs perfluoronated gross leak testing. The vessel is then pressurized to a given pressure. This will force the Type 1 fluid into the cavity of the product. After a given period of time, the product is removed and put into a heated tank of Type 2 fluid and typically heated to 125°C. Type 1 fluid will boil at 95°C, and Type 2 fluid will boil at 140°C. By observing a device under test in the tank at elevated temperature, the Type 1 fluid can be observed to boil off in the presence of a gross leak. This will appear as bubbles in the Type 2 fluid, which is kept below its boiling point. This method is very useful in determining leaks and locations of anomalies occurring from leaks.

The penetrant dye gross leak method is useful in determining the physical structure and/or location of a leak. The procedure requires that the device under test be placed in a pressurized bath of dye such as Zyglo or fluorescence. While the device under test is under pressure, the dye will penetrate any defects in the surface of the unit and enter the device cavity. After a given period of time, the unit is removed from the vessel, rinsed with the appropriate solvent to remove the surface dye, and placed under a black light. As the dye fluoresces, the physical characteristics of the leak and leak path can be observed. Since this is a destructive test, it is used for monitoring rather than for screening products.

5.3.6 Particle Impact Noise Detection (PIND)

Particle impact noise detection is used to detect solder balls and other foreign substances in hermetically sealed packages. The unit is attached to the top of a transducer, vibrated at a given frequency and amplitude, and shocked at a given pulse to loosen any particles. The signal from the transducer is monitored to detect acoustic signals from loose particles with results displayed on an oscilloscope. Any excess signal ringing or spike is a sign of foreign particles inside the package and is a cause for device rejection.

There can be a few problems with the test itself. For example, the device under test may have a construction that can cause a false reading due to loose ferrite beads, transformer material, coils, wireloops, etc. This is a false reading since test results are unrelated to loose particle problems. Additional test problems can occur due to lack of proper electrical detector isolation from stray signal noise that commonly occurs in switches, electrical lines, mainstream factory signals, and so forth. Test results can be questionable.

5.3.7 Ionic Contamination

Ionic contamination testing is used mainly to monitor solder flux cleaning processes. A sample of board-level assemblies is taken from the manufacturing line, just after the final cleaning and before the next assembly step. These boards are placed in an agitated bath of alcohol and water with a known level of resistance. If the boards have any flux or contaminant left on them that is soluble in water or alcohol, it will decrease the resistance of the bath fluid. This indicates that the cleaning process for the lot is not sufficient or that the cleaning fluid needs to be changed. Care must be taken to ensure that the boards are handled properly (with gloves, etc.) after cleaning, or the boards may falsely fail. The acceptance level is based on the resistance per square inch of total surface area of the boards.

5.3.8 Environmental Stress Screening (ESS)

Traditional ESS is performed on 100 percent of the production units. Typical ESS accelerated environments that already have been specified include:

1. *Operating heat soak (sometimes called burn-in):* This is accomplished by operating the units at a warm constant temperature for a given time period. A typical example is 48 hours at 50°C.

2. *Thermal cycling (this may be powered or unpowered):* This is accomplished by running a stated number of cycles in which each cycle contains a fixed time period at a high temperature, a fixed period at a cold temperature, and a thermal transition rate between the extremes measured in °C/minute.

3. *Fixed-frequency sine vibration:* This is used either alone or with some other environment. The ESS level is defined by the amplitude and vibration frequency at the equipment mounting point.

4. *Swept sine vibration:* This is occasionally used for ESS. The ESS level is defined by gravitational (G) levels over a specified frequency range with a specific sweep time.

5. *Random vibration:* This is also used for ESS. The ESS level is defined by the vibration amplitude level and frequency spectrum at the equipment mounting point.

5.3.9 Vibration/Constant Acceleration

Constant acceleration, as the name indicates, is a test that exposes a product to a constant gravitational force in one direction. The test uses centrifugal force to obtain very high gravitational levels (usually 5,000 to 20,000). It can simulate how a part or system will react to the effects of constant accelerated stress in aircraft, missiles, etc. As a screen, it is designed to indicate the types of structural and mechanical weaknesses not necessarily detected in vibration and mechanical shock. By establishing a nominal level, it may be used to detect and eliminate devices with lower-than-nominal mechanical strengths in any of the mechanical structures.

5.3.10 Humidity

Humidity tests accelerate the effects that moisture can cause in components. Degradation results if materials have problems with moisture absorption and/or surface wetting. Problems include corrosion, changes in electrical properties, electrochemical reactions, and so forth.

Two types of test procedures can be performed. The first is to expose parts to steady-state humidity at an elevated temperature over time. Chapter 9 describes an accelerated testing model that may be used to estimate the accelerated effects of the test relative to field-use conditions. This test is usually accompanied by power cycling in which bias is turned on and off in four-hour increments. This promotes any dendrite growth or bias corrosion effects due to different ionic contaminants in the material that may be present from the manufacturing process. During the off power cycle, moisture is able to reach components that dissipate too much heat when powered. The test also is used to test the well-known "popcorn" effect in plastic encapsulants. This problem occurs in an encapsulant that absorbs excess moisture. During a solder reflow process in manufacturing, the plastic or housing expands rapidly and ruptures, exploding like popcorn due to the excess absorbed moisture.

An alternative humidity test procedure is to place components in a chamber that is temperature cycled while the humidity is elevated and held constant. This promotes a "breathing" effect that allows moisture to work into the small cracks and fissures. At the end of the 24-hour cycle, the temperature is dropped to −10°C or −20°C. Any moisture in the material will freeze and expand, accelerating the process of fissures and cracks. The results are typically degradation in electrical performance and voltage breakdown from lowered insulation resistance.

5.3.11 Highly Accelerated Stress Test (HAST)

Highly Accelerated Stress Test (HAST) produces the same effect as the steady-state humidity test described above but in a shorter time period. This is accomplished by introducing higher temperatures (>100°C) in a humidity chamber than in the steady-state test chamber (which is usually at 85°C). These higher temperatures can be obtained in a special sealed HAST chamber. The sealed chamber allows for higher-than-atmospheric pressure, enabling control of above-average temperatures (greater than the 100°C boiling point of water) in a controlled-humidity environment.

Testing with a typical humidity at 85°C and 85 percent relative humidity (RH) steady-state atmosphere for 1000 hours (roughly representing a 10-year life span at 40 percent RH and 50°C typical use environment) can be performed in a HAST test (at 17.6 psia) for 135 hours at 120°C and 85 percent RH. This assumes the model and parameters described in Chapter 9 (see Example 9.3). If the device temperature exceeds the chamber ambient by more than 2°C, or if the dissipation of the device exceeds about one-tenth of a watt, device bias should be cycled with a 50 percent duty cycle. This allows the local relative humidity of the device to reach chamber ambient at least 50 percent of the time; otherwise, the test effects will be negated.

5.4 Highly Accelerated Stress Screening (HASS)

The information obtained when a product is first introduced to the Development Phase in the HALT process (see Chapter 3) enables the development of a HASS test. HALT, as described in Chapter 3, is a highly accelerated reliability growth Test-Analyze-And-Fix (TAAF) process.

In HASS, failures are analyzed, and corrective actions are implemented. The test is repeated until the observed failure modes have been fixed, and the environmental technology limits of the part are understood. This information is used for the Production Phase.

At this stage, one either develops a traditional screen or a HASS test. The traditional screen will employ one or more of the tools described above to look for latent defects. The next option is to develop a HASS test that combines thermal cycling, vibration, and power stress simultaneously. The testing range is within the operating limits that are known from prior HALT testing performed in stage gate 2.

Similar to HALT, HASS is an aggressive screening program to help weed out failure modes and implement corrective actions as soon as possible. This process enables products to be moved quickly into a monitoring program. This entire process is depicted in Figures 3.3 and 3.4.

The HASS process typically helps to reduce screen time (30 percent to 80 percent) and move a product more quickly into the Monitoring Phase. For example, a common screen uses 168-hour burn-in, 20-hour thermal shock, and a 60-minute vibration test. Since this is a fairly lengthy screen, it is advantageous to work with a HASS program. In the HASS process, this test is quickly reduced to a monitoring program. Since faster test results help in implementing product improvements and moving to a monitoring test, cost savings can be passed onto the customer.

If HASS precipitates a failure, an immediate failure analysis is performed to determine the root cause. A 100 percent screen is maintained until the process is in control. At that point, monitoring can be performed. The monitoring also includes a HALT at given intervals to ensure that the product safety margins have not deteriorated from those obtained in the Evaluation Phase. Thus, knowledge of the HALT and HASS environmental limits, relative to a customer specification, is very helpful in providing engineering confidence in the proper design of the screening and the subsequent monitoring test. Such sound practices are important for providing a highly reliable product.

SECTION II

SUPPORTING STAGE GATE

CHAPTER 6

Semiconductor Process Reliability

6.1 Introduction

Semiconductor process reliability studies are used to quantify the wearout characteristics of semiconductor devices. When assessing the reliability of a semiconductor product, there are many elements that need to be considered. A typical product consists of a semiconductor die attached to a package using either solder or epoxy, with wire bonds between the die bond pads and the package pins. Problems with any of these elements, either individually or in interaction with other parts of the assembly, may lead to device failure. In order to systematically measure the contribution of any of these elements to product reliability, it is often useful to design reliability tests that stress only one element of the entire assembly. The semiconductor die is usually the most complex single element of the product assembly and is therefore the element most susceptible to wearout and eventual failure.

The objective of performing semiconductor process reliability studies is to assure that the semiconductor device manufacturing processes are capable of producing products with acceptable long-term reliability. This objective is typically achieved by performing a series of experiments that stress test vehicles manufactured using these processes over a range of environmental conditions. Functional dependence of device degradation to these stresses is monitored during testing. Appropriate stresses must be used so those devices either degrade parametrically or fail catastrophically. From these tests, it is possible to identify and quantify the reliability-limiting mechanisms in the semiconductor manufacturing process.

Statistical analysis of the experimental data allows a baseline model for semiconductor process reliability to be obtained. This model can be as simple as estimating device performance under a known set of environmental conditions or as complicated as a statistical reliability predictive model. In addition, the major failure mechanisms acting to degrade the electrical performance of the test vehicle can be experimentally identified, and the functional degradation dependence caused by the environmental stress determined. Process and design engineers may then use this information to improve the processes with respect to the failure mechanisms, resulting in higher reliability. The ideal steps in a semiconductor process reliability study are:

1. Identify and quantify the major failure mechanisms acting on devices manufactured with some set of processes.

2. Establish the functional dependence of reliability on device application conditions.

3. Determine a baseline model for semiconductor process reliability.

4. Feed these results back to the design and process engineering disciplines in an effort to improve product reliability.

In this chapter, the discussion will be limited to processes employed in the manufacture of semiconductor dies. The specific focus will be on semiconductor process reliability techniques for Gallium Arsenide technologies. These same basic techniques can be generally applied to all types of semiconductor devices.

6.2 Overview of Semiconductor Process Reliability Studies in the GaAs Industry

This section summarizes the major techniques currently used across the GaAs industry to study process reliability. An overview of the failure mechanisms identified during these experiments will be provided. This overview will provide a basis for the proposed approach to semiconductor process reliability testing.

6.2.1 Test Vehicles

In defining a semiconductor process reliability study, the first requirement is often to determine what to use as a test vehicle. The three types of test vehicles that may be employed are integrated circuits (ICs), discrete components (FETs, capacitors, resistors, etc.), and process-specific test structures. Each of these carries associated advantages and difficulties.

Perhaps the most frequently used test vehicle is the IC. For GaAs devices, this is more commonly known as a monolithic microwave integrated circuit (MMIC). This is often an actual product, but it may also be a circuit designed specifically to exercise all of the discrete elements of the process. Using an actual product as a reliability test structure usually means that fixturing and test requirements have already been addressed for manufacturing purposes and that an ample supply of devices is readily available. Another advantage is the direct applicability of the results of reliability studies on a product to that same product or product family. The major difficulty with using an MMIC is that electrical, statistical, and physical failure analyses can be complicated, since the test vehicle is more complex (see Figure 6.1) than discrete elements.

Discrete components that are used in the design of ICs may also be used for semiconductor process reliability studies. Studies of this nature have been published elsewhere (see Reference 1). Using these as test vehicles, it is easy to identify the failure mechanisms since each discrete element can be examined and characterized electrically. Once characterized with respect to reliability, any new product that utilizes these components can be modeled and assessed at the design stage. An example of a discrete component test vehicle includes a transistor (MESFET), as well as capacitors and resistors, which can be tested and characterized in a reliability study (see Figure 6.2).

The three major drawbacks to the use of discrete components as test vehicles for semiconductor process reliability studies are fixturing, availability, and correlation. It can be challenging to design fixtures that survive high-temperature reliability experiments and still provide a stable biasing environment for the test vehicle. This challenge increases for high-frequency, high-gain, and submicron gate-length GaAs FET applications. These devices are extremely susceptible to bias oscillations. Component availability may also lead to problems in obtaining sufficient quantities for performing statistically valid life tests based on discrete elements. This is typically a problem when components are not manufactured as products. The ability to model circuit reliability using process reliability results can be

Figure 6.1

Example of a low-noise MMIC amplifier used as a reliability test vehicle

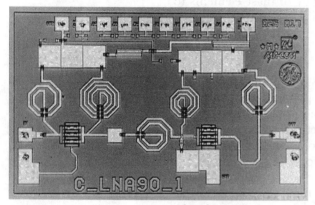

Figure 6.2

Device designed specifically as a reliability test vehicle

difficult. It is important that the reliability models correlate closely to what is observed when life tests are performed on actual products.

Process-specific test structures share the same advantages and disadvantages as discrete components in this application. Failure analysis of these structures can be reasonably straightforward, as they are designed specifically to characterize a particular process. Some of these structures include:

- Transmission Line Model (TLM) devices (resistor ladders), used to characterize contact and sheet resistance.
- Van der Pauw crosses, used to characterize sheet resistance and structure critical dimensions.
- Air bridge chains (or chains of other structures), used to investigate process defect densities.
- Parallel lines of various widths and spacing, used to investigate current-dependent electromigration and field-dependent shorting.
- Metal-Insulator-Metal (MIM) capacitors, used to investigate the quality of dielectric films (Si_3N_4, SiO_2).
- Fat FETs (FETs with large gate-lengths 50–100 μm), used to investigate Schottky diode characteristics, doping, and mobility profiles.

There are a number of different types of process-specific test structures (see Figure 6.3). Each of these elements may contribute to a reliability model; however, the potential problems with fixturing, availability, and correlation discussed previously still apply.

Figure 6.3
Process control monitor, containing several structures which can be used as reliability test vehicles

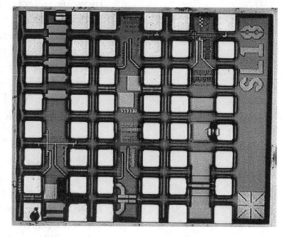

6.2.2 Experimental Design

Semiconductor process reliability studies typically use significantly smaller sample sizes than product qualifications. When the reliability of a process is first being investigated, sample sizes of ten or fewer devices may be used in step-stress experiments to obtain a quick, approximate measure of the process sensitivity to various forms of accelerated stresses. Temperature is the most common stress applied, as it affects most of the major GaAs wearout failure mechanisms. A typical step-stress test may consist of ten devices stressed for 24-hour periods, starting at 125°C and increasing the temperature by 25°C each day until at least 50% of the population fails (often 6 or more steps). The test vehicles are electrically characterized before and after each step. Units may or may not be biased during the stress intervals. From the step-stress test results, a qualitative decision can be made regarding stress levels to apply during constant stress life tests.

With step-stress test results and/or previous experience with the process in question, a constant stress reliability study can be designed. Typical sample sizes during these tests range from 10 to 40 devices per stress condition, with at least two, and preferably three or more, combinations of bias and temperature conditions applied. Channel temperatures from about 125°C to 300°C are commonly used. For small-signal (linear) applications, devices are typically biased only with direct current (DC). For applications where devices operate under conditions of significant gain compression (nonlinear conditions), it is desirable to perform life tests under RF-biased conditions. This is because the DC-bias approximation to the actual application environment is less valid. Sample sizes during RF-biased tests, compared with DC-biased life tests, are typically smaller, simply because the cost per sample for fixtures and for the life test system is so much greater.

While temperature is the most commonly used acceleration factor, other stresses may also be important to fully understand the reliability characteristics of a process. Current density and voltage are bias-related acceleration factors that may have a significant effect on GaAs devices; however, test vehicles are generally not stressed over a range of current densities or voltages. Devices are usually biased at a nominal voltage and current level while subjected to elevated temperatures. This approach, which is valid for addressing all products manufactured with given processes that use essentially the same bias conditions, greatly simplifies the studies.

6.2.3 Life Test Execution

Once the bias and temperature conditions for constant stress testing have been determined, the accelerated life tests may begin. As with step-stress tests, thorough electrical characterization is performed for each test device before, periodically during, and after stress. These measurements are made at room temperature. In addition to these measurements, the bias conditions (currents and voltages) of each device are monitored during the stress periods. The evolution of device electrical performance as a function of time under stress is analyzed to determine when each part fails, either catastrophically or parametrically. The life test typically continues until least 50 percent of the devices have failed.

6.2.4 Data Analysis

Analysis of the data collected during constant stress life tests is performed at three levels: electrical behavior of each device, statistical behavior of all the devices stressed at a single condition, and statistical behavior of all the stress levels included in a semiconductor process reliability study. These analyses, combined with physical failure analysis, provide an understanding of the failure mechanisms exhibited by a process and the functional dependence of those mechanisms on the environmental conditions (stresses) applied during the life tests.

Electrical data analysis may also include *in situ* (monitored) measurements. Monitored data typically consist of bias voltages and currents, and/or parameters calculated from these bias conditions. Figure 6.4 shows the monitored values of beta (DC current gain) for 11 bipolar transistors during a life test. These data allow the specific time at which failure occurs to be determined with great accuracy for each device in the life test population. For this specific test, failure was defined as a degradation of greater than 10 percent in beta. Replotting the data (see Figure 6.4) relative to the initial measured value of beta provides a simpler means of determining the specific time to failure for these devices (see Figure 6.5).

Not all device parameters can be easily monitored during the life test. This is especially true for RF and microwave parameters. It is therefore important to periodically interrupt the application of stress and electrically characterize the devices at room temperature. Room-temperature measurements also provide a means of checking whether the rate of degradation observed at high temperatures correlates to performance degradation at application temperatures. An example of this kind of measurement is shown in Figure 6.6, which shows the change in gain at 9 GHz for X-band low-noise amplifiers (LNAs) as a function of time under DC-biased stress at an FET channel temperature of 200°C. As with the monitored data, these room-temperature measurements may be used to determine the time to failure for the devices in a life test.

An analysis of these measurements, coupled with an understanding of how physical changes in the device would affect electrical performance, is often important for identifying the specific failure mechanisms observed.

As device failures begin to occur, the second level of data analysis can begin. The objective of this analysis is to statistically summarize the results of a life test. Determining a mathematical function that accurately describes the dis-

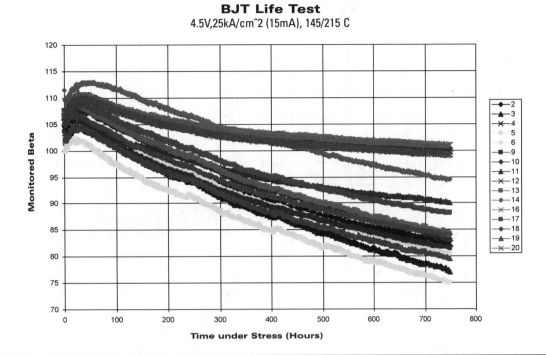

Figure 6.4
*Beta monitored during
a life test of bipolar transistors*

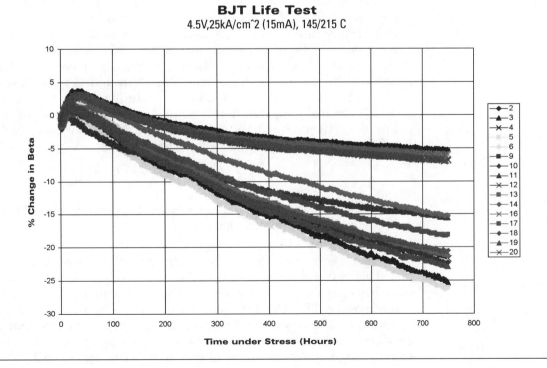

Figure 6.5
*Percent change in
beta monitored during
a bipolar transistor life test*

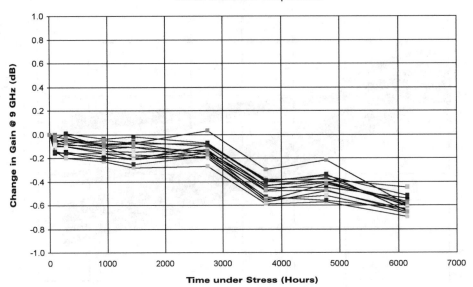

SEC DC Biased Life Test
at 200°C Channel Temperature

Change in Gain @ 9 GHz (dB) vs *Time under Stress (Hours)*

Figure 6.6

Change in gain as a function of time under stress for LNA test vehicles

tribution of device failures in time, for each unique life test condition, does this. Commonly used distributions for microelectronics reliability studies include the log-normal (see Figure 6.7) and the Weibull distributions. The two curves shown in Figure 6.7 are composed of the times to failure of X-band LNA devices life-tested under DC bias at 200°C and 250°C. The plot also displays the median time to failure, the log-standard deviation value sigma, and a goodness-of-fit coefficient for each set of failure data.

Once failure distributions that describe the reliability characteristics of a process under various sets of environmental conditions have been determined, the final level of data analysis can begin. This involves modeling the functional dependence of device reliability characteristics (i.e., lifetime) on the level of stress the device was subjected to during the life test. For life tests in which elevated temperature is used to accelerate device degradation and failure, the Arrhenius model is typically applied. As it applies to reliability testing, the Arrhenius model can be written as:

$$MTTF = Aexp(E_a/kT) \qquad (6.1)$$

where

$MTTF$ = the mean-time-to-failure at some temperature T (in Kelvin),
E_a = the activation energy, and
A = a constant.

A plot of one over the life test temperature against reliability data is often presented in terms of the instantaneous failure rate, which is also called a hazard rate (see Chapter 8). While MTTF is specified with units of hours, the hazard rate uses a unit known as FITs. One FIT is equal to one device failure in 10^9 device-hours of operation. The hazard rate at any temperature can be calculated from the failure distribution parameters at that temperature. For an example of a hazard rate plot for a log-normally distributed failure mechanism, see Figure 6.9.

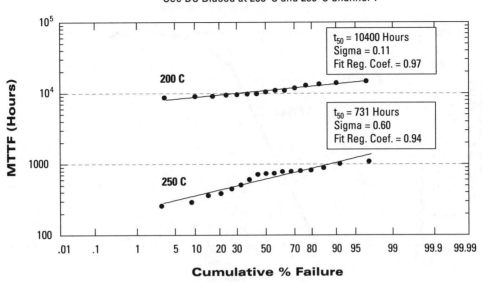

Figure 6.7
*Log-normal failure
distributions for LNA
devices life-tested at
200°C and 250°C*

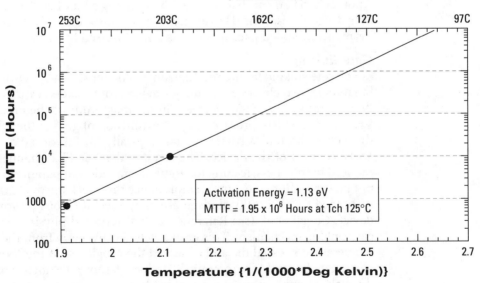

Figure 6.8
*Arrhenius plot from a DC
life test on MMIC LNAs*

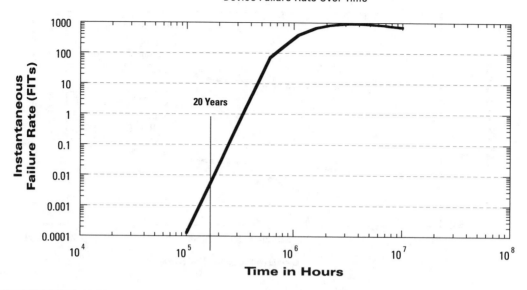

Figure 6.9

Example of a hazard rate plot

6.2.5 Observed Failure Mechanisms

When developing a process reliability study, it is important to consider the types of failures that might occur. To help understand how an MESFET is constructed and where failure mechanisms could physically act upon the device, a cross-section of the FET used in a 1μm switch process is shown in Figure 6.10.

The most commonly reported failure mechanisms for GaAs MMICs in recent years include gate sinking and hot electron trapping. Other mechanisms such as ohmic contact degradation and surface-related (passivation) degradation, while common in the early to mid-1980s, are infrequent today. Another mechanism that is currently receiving a lot of attention is referred to as hydrogen poisoning. However, this mechanism only affects GaAs devices in hermetically sealed packages and is not addressed here.

Gate Sinking

The gate sinking failure mechanism can be most easily identified by a combination of two electrical parameter degradations: a decrease in the saturated drain current, I_{DSS}, and a decrease in the magnitude of the pinch-off voltage, V_P. The mechanism proceeds with the diffusion of gold in the FET gate, either through or around a barrier metal (typically Pt, Pd, or TiW), and into the active channel of the device. The result is that the FET channel is effectively reduced in thickness, leading to circuit degradation in parameters such as output power and/or gain. Furthermore, since this mechanism degrades the quality of the FET's Schottky diode, it also may affect the noise performance of the device. Canali et al. (see Reference 2) first reported physical evidence of the gate sinking mechanism. This group etched GaAs away from the backside of a temperature-stressed die and observed the roughness at the "bottom" surface of the gate finger. More recently, Roesch et al. showed evidence of gate sinking through the application of focused ion beam cross-sections to gate fingers (see Reference 3). The rate of degradation due to this mechanism is accelerated by elevated temperature, with activation energies reported between about 1 and 2.5 eV. While there is some anecdotal evidence that this mechanism is affected by device bias conditions, there have been no published studies relating the rate of degradation to bias during accelerated life tests.

Hot Electron Effects

Recent studies reported by Hwang et al. (see Reference 4) suggest that hot electron effects lead to GaAs MESFET degradation and failure, particularly for devices that operate under highly nonlinear conditions. It has been proposed that, as the potential field gradient in the MESFET channel increases, enough energy may be transferred to electrons to allow some percentage of them to become trapped in the Si_3N_4 passivation layer between the gate and drain fingers (the region of highest potential field gradient). These trapped electrons create a depletion region between the FET gate and drain fingers which is not modulated by gate bias. The result is effectively a permanent constriction in the FET channel, leading to decreased I_{DSS} and output power. This mechanism is most significant for saturated power amplifier applications, as the FET must be biased heavily into the three-terminal breakdown region before the potential field gradients are sufficient to cause this mechanism to occur.

Passivation Degradation

Several papers discussing this mechanism were published by Dumas et al. in the early to mid-1980s (see Reference 5). This mechanism is typically observed by its impact on gate leakage currents. The interface between GaAs and as-deposited silicon nitride (Si_3N_4), a commonly used device passivation material, may be quite poor due to mechanical stresses at the interface. This results in a high density of surface states, which can serve as gate current leakage paths. It is often observed that, with temperature stress, this leakage current will initially decrease as the GaAs/Si_3N_4 interface is annealed. After some length of time under stress, however, the leakage current begins to increase, suggesting a competing mechanism that degrades the interface (see References 6 and 7).

Ohmic Contact Degradation

This failure mechanism was commonly reported from reliability studies performed in the early and mid-1980s, but seems to be much less prevalent today. Ohmic contacts to N^+ GaAs are usually formed with AuGe. Advances in annealing processes over the past several years have minimized the impact of this mechanism for contacts to N-type GaAs. However, there may still be issues with ohmic contacts to P-type GaAs.

Figure 6.10

Cross-section of a MESFET

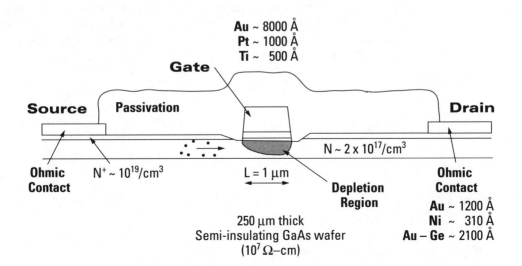

6.3 Wafer Level Reliability Tests

Among the most significant semiconductor process reliability developments in recent years are tests designed to gauge reliability at the wafer level. The primary intent of these tests is to provide a real-time monitor of process reliability for each wafer that is manufactured. Several advantages are realized through the use of these tests. These tests are designed to run very quickly, typically requiring only seconds per sample. As a result, reliability information for each manufacturing lot is available before products are delivered to customers. Since these tests take little time to perform, it is possible to collect data from hundreds of samples. With data from so many devices collected across multiple manufacturing lots, a statistically significant measure of reliability variation within a process can easily be obtained. The rapid nature of these tests facilitates their use for quickly analyzing the impact of process changes on device reliability, allowing rapid deployment of new processes and products to market.

The wafer-level reliability tests available today have been developed in the silicon device industry to address a number of the dominant failure mechanisms in their technologies. Tests have been reported for electromigration, time-dependent dielectric breakdown, and hot (highly accelerated) carrier effects in MOS technologies. Unfortunately, these are not the primary failure mechanisms reported in GaAs; therefore, these techniques are of limited use for GaAs processes. The major wearout failure mechanism reported for GaAs MESFET devices, gate sinking, does not seem to readily lend itself to wafer-level characterization. Only one approach to highly accelerated stress testing for a GaAs device has been reported (see Reference 8). This technique seems to have received little attention in the industry. Until some revolutionary technique is developed, it appears that process reliability studies addressing wearout mechanisms of GaAs MESFETs will continue to be performed in the traditional, time-consuming way.

6.4 Summary

This chapter has presented an approach to semiconductor process reliability testing, focusing on GaAs technology. Several reasons for performing these tests were discussed, including identifying the major failure mechanisms active in a process and establishing the functional dependence of these mechanisms on device application conditions. This information is used to establish a baseline model for semiconductor process reliability, from which the reliability impact of process changes can be assessed. Finally, the ultimate objective of these tests is to assure that a process is capable of producing highly reliable products. However, it is only through the combined application of additional reliability techniques that reliable products may be realized.

References

1. Dumas et al., "Comparative Reliability Study of GaAs Power MESFETs: Mechanisms for Surface-Induced Degradation and a Reliable Solution," *Electronics Letters*, Vol. 21, pp. 115-116, 1985.

2. Canali et al., "Gate Metallization Sinking into the Active Channel in Ti/W/Au Metallized Power MESFETs," *IEEE Electron Device Letters*, March 1986.

3. Roesch et al., "Life-Testing GaAs MMICs under RF Stimulus," *IEEE Transactions on Microwave Theory and Techniques*, December 1992.

4. Hwang, "Gradual Degradation under RF Overdrive of MESFETs and PHEMTs," *IEEE GaAs IC Symposium Proceedings*, 1995.

5. Dumas et al., "Long-Term Degradation of GaAs Power MESFETs Induced by Surface Effects," *IEEE International Reliability Physics Symposium Proceedings*, 1983.

6. Ersland et al., "GaAs FET Switch MMIC Reliability," *IEEE GaAs IC Symposium Proceedings*, 1988.

7. Ersland et al., "GaAs FET Switch MMIC Reliability Revisited," *IEEE GaAs Reliability Workshop Proceedings*, 1990.

8. Yeats et al., "Method for Rapid Determination of Activation Energy and Lifetime from Accelerated Stress of a Single Device: Theory and Experiment," *IEEE GaAs Reliability Workshop Proceedings*, 1992.

CHAPTER 7

Analytical Physics

7.1 Introduction

Physical analysis, also known as analytical diagnostics or failure analysis, is a major element of reliability engineering. This type of analysis will answer most of the **why, where, when,** and **how** questions about the life of a component. Analysis methods prove essential in understanding, determining, and applying appropriate corrective action for root cause of failure. Understanding the root cause of a failure is essential in today's highly competitive market for successfully manufacturing quality components. In this chapter, we discuss the basic methods in performing a root-cause physical analysis and present an overview of analytical techniques for analyses.

7.2 Physics of Failure

Determining the physics of failure after a controlled experiment or after a field failure is essential in understanding the products and their limitations. Finding a root cause is beneficial to decrease repeat failures and set the specification limits on products. Customer satisfaction often depends upon resolving problems and preventive actions that eliminate future anomalies.

Diffusion, corrosion, dendritic growth, contamination, effects of stress-strain, and ESD are major factors when determining the root cause for a failure. In the next few sections, these topics are discussed and are related to real-life situations.

7.2.1 Diffusion

Diffusion is defined as the migration of atoms or mass transport. It is well known that atoms in gases and liquids are very mobile. For example, when a bottle of perfume is opened, it can be smelled in adjacent rooms quickly, sometimes almost immediately. It is not well known whether atoms in solids also are mobile and can move readily from one area to another. This phenomenon is used in the production of quarters and galvanized steel. For the case of the quarters, a metal sandwich is made with a copper/nickel alloy surrounding a copper center. The copper diffuses both ways and produces a metallurgical bond. A detrimental aspect of diffusion is Kirkendall voiding, where one atom in a metal-to-metal joint diffuses so quickly that vacancies (holes in the lattice) accumulate at the backside of the diffusion line and weaken the material.

There are two types of diffusion in solids: interstitial and substitutional, or vacancy. Interstitial diffusion is primarily the diffusion of the light elements hydrogen (H), carbon (C), and nitrogen (N). This is the most rapid type of diffusion because these smaller atoms move between interstitial sites within the lattices since these sites are usually empty. An interstitial site is a space in a lattice between the primary lattice sites.

Substitutional diffusion, or vacancy diffusion, is slower than interstitial diffusion because the atoms move from vacancy to vacancy, and the diffusion process must wait to move until a vacancy opens up in the lattice. A vacancy is a point defect or a missing atom in a lattice. The number of vacancies in a material increases with temperature because the vibration of the atoms increases and more defects are created. This is why diffusion happens more readily at higher temperatures. This can also be shown by the following equation:

$$D = D_0 e^{-Q/(RT)} \qquad (7.1)$$

where

D = the diffusivity or diffusion coefficient,

D_o = a material constant,
Q = the activation energy,
R = the gas constant, and
T = the temperature in degrees Kelvin.

Diffusion is also time-dependent, shown by the following equation:

$$\frac{C_s - C_x}{C_s - C_o} = erf\left(\frac{x}{2\sqrt{Dt}}\right) \qquad (7.2)$$

where
  t = the time,
  C_s = the surface concentration,
  C_x = the concentration at some distance x, and
  C_o = the original concentration.

Furthermore, diffusion is gradient-dependent. It is identical to a temperature gradient where the heat will go from hot to cold until equilibrium is reached. If the inside of a house is 20°C and the outside is 0°C, the heat from the inside will dissipate to the outside. Diffusion of solids works the same way. The following equation shows the effect of the concentration gradient:

$$J = -D\frac{dc}{dx} \qquad (7.3)$$

where
  J = the flux,
  D = the diffusivity, and the derivative yields the concentration gradient.

The Kirkendall Effect is the occurrence in diffusion when metal A and metal B are placed together, and metal A diffuses into metal B faster than metal B diffuses into metal A. After some time, the interface between the two metals will move toward metal A:

Figure 7.1

These x-ray maps show the Kirkendall effect. The virgin sample (left) has clear delineated layers (Ni, Au, Sn). Over time and temperature, the interface of the gold (Au) and tin (Sn) moves (right).

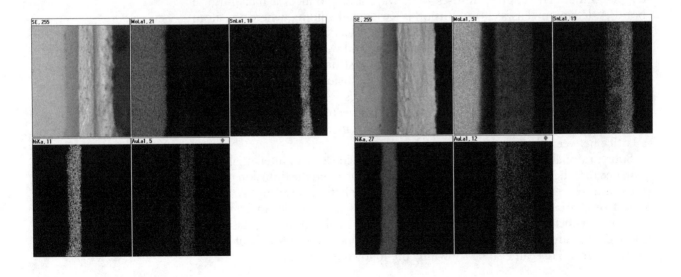

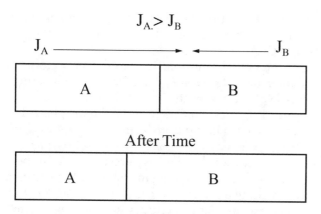

$$J_A > J_B$$

After Time

If the diffusion of metal A atoms into metal B atoms is faster than the A atoms can fill their own vacancies, a voiding will occur within metal A or at the interface (Kirkendall voiding):

$$J_A > J_B \Rightarrow J_{V_A} < J_{V_B} \qquad (7.4)$$

J_A, J_B = flux of atoms J_V = flux of vacancies

7.2.2 Phase Diagrams

One of the most useful tools in materials science for understanding multiple element systems is the phase diagram. It is a graph that shows the phase or phases present at different compositions as a function of temperature under fixed pressure (normally 1 atmosphere). They can also show solubility and quantitatively determine how much of a phase is present (Lever Rule). The following phase diagrams show a system that forms a solid solution (A is soluble in B), and a system that has a eutectic point.

There are two types of solid solutions: substitutional and interstitial. An

Figure 7.2
The SEM micrograph shows Kirkendall voiding. The Au layers diffuse rapidly into the InP layer and produce a crack at the original interface.

Figure 7.3
The phase diagram at the left shows a solid solution while the phase diagram on the right shows a eutectic point.

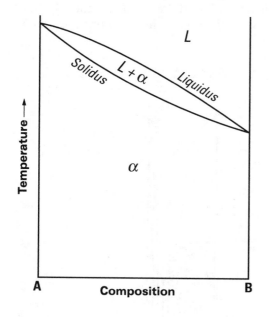

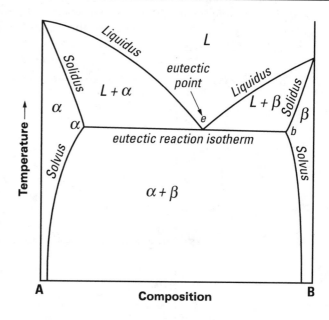

Figure 7.4 (below)
Substitutional solid solution of Ni in Cu

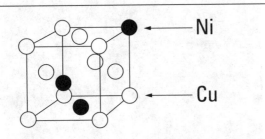

example of a substitutional solid solution is nickel (Ni) in copper (Cu). When Ni forms an alloy with Cu, the Ni atoms substitute for the Cu atoms in the basic lattice structure (FCC).

An interstitial solid solution has the alloy atoms in the interstices. These are primarily the smaller atoms because of the size of the interstitial holes.

Eutectic is Greek for "melts well." On the phase diagram, it is when liquid transforms to two solids upon cooling (a eutectoid is solid to two solids). At the eutectic point, an alloy has its lowest melting temperature. This allows the alloying of metals to have a lower melting point than those metals alone. A eutectic is also a lamellar structure, where alternating layers of the constituents comprise the phase.

Figure 7.5 (below)
This micrograph shows a eutectic microstructure

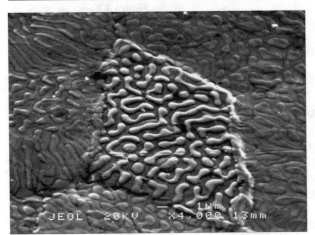

For example, components are soldered onto a board using a noneutectic solder. This board is then soldered onto another board that is in its final housing. The solder used for the final placement is eutectic. This allows the heating and melting of the eutectic solder to place the final component without reflowing the original solder on the first board. Phase diagrams can ascertain these temperatures.

7.2.3 Intermetallics

Intermetallics are metal-to-metal phases such as AuSn and AuAl. The aforementioned systems are well known for the presence of gold embrittlement and purple plague. It is important to note that intermetallics are not always bad. In the case of the Au/Al system, there are five AuAl intermetallics that are formed, and all are electrically conductive and mechanically strong. These can be found in the phase diagram in Figure 7.6. The problem exists because of the unbalanced diffusion rates where

Figure 7.6 (below)
Au/Al phase diagram (from Reference 1, with permission)

Kirkendall voiding may occur. A second phenomenon of refining occurs when a zone of contamination is pushed in front of the diffusion line. Eventually the concentration of the impurities increases and weakens the bond or makes the joint

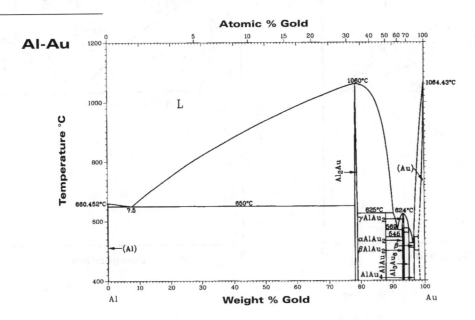

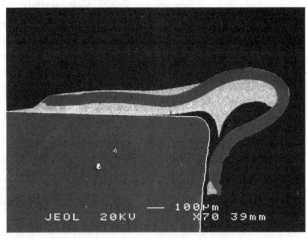

Figure 7.7
AuAl$_2$ intermetallic growth and Kirkendall voiding of a ball bond (from Reference 2, with permission)

Figure 7.8
Typical Au embrittlement solder fracture

electrically resistive. Thermal shock and temperature enhance the condition of purple plague (as stated previously where diffusion is temperature-dependent). Gold embrittlement is the formation of AuSn$_4$ which is a brittle intermetallic that forms in the Au/Sn system from 4 wt % Au to 43 wt % Au. The AuSn$_4$ is distributed as plates and needles within the matrix. This formation mostly occurs in solder joints that use high Sn solder and thick Au layers. In the microelectronics industry, the Au top plating is used to prevent oxidation or corrosion of the adhesive/diffusion barrier layer underneath (many times Ni). The Au plating dissolves into the molten solder quickly and, if the percentage is high enough, forms AuSn$_4$. A plating thickness of less than 40 microinches should be used to keep the weight percentage of Au below 4% (less, if only a small volume of solder can be used). Figure 7.8 shows a typical Au-embrittled fracture of a soldered lead.

The percentage of Au in the solder will affect the mechanical properties greatly. It can be seen in Figure 7.9 that a Au percentage between 5% and 6% will decrease the mechanical properties over 80%. Figure 7.10 shows an x-ray map of a Au-embrittled solder joint cross-section.

Figure 7.9
Impact toughness vs. Au content (see Reference 3)

Figure 7.10
X-ray map of Au-embrittled Sn62 solder joint

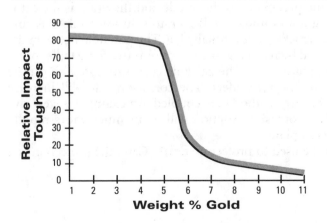

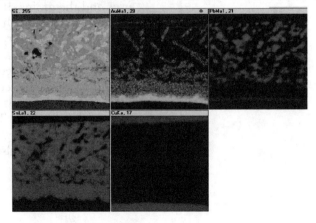

7.2.4 Corrosion

Corrosion is the deterioration or destruction of a material because of the reaction with its environment. Every year, the U.S. spends over $50 billion on corrosion, from bridges and decks that are driven on to the cars that are driven in and water heaters in houses. Corrosion has four basic requirements. Without one of these requirements, there will be no corrosion. These are:

1. Anode – Chemical oxidation

$$Fe \rightarrow Fe^{2+} + 2e \qquad (7.5)$$

$$M \rightarrow M^+ + ne \qquad (7.6)$$

2. Cathode – Chemical reduction

$$O_2 + 2H_2O + 4e \rightarrow 4OH^- \qquad (7.7)$$

3. Electrolyte – liquid and mobile ions
4. Conductive path

The factors affecting corrosion rate are the material properties such as the grain size, impurities, and the electromotive force. Environmental conditions such as humidity, temperature, electrical bias, and ionic contamination also greatly affect the corrosion rate. There are eight forms of corrosion:

Uniform – general attack
Galvanic – two-metal or battery type
Crevice corrosion – localized corrosion within a crevice
Pitting – extremely localized attack leaving pits in the material
Intergranular – localized attack at and near grain boundaries
Selective leaching – removal of one element of a solid solution by corrosion
Erosion – acceleration of corrosion because of wear and abrasion
Stress corrosion – failure due to corrosion and a tensile stress

Only uniform attack, galvanic corrosion, selective leaching, and stress corrosion will be discussed here.

Uniform attack is the most common form of corrosion, where there is a chemical reaction over the entire surface of the exposed area. Here, the metal will thin until failure. The best way to decrease uniform attack is to paint or coat the part to make sure it does not get wet.

Galvanic corrosion is when two dissimilar metals have a potential between them producing electron flow. Higher potentials between metals cause increased electron flow and faster corrosion rates. Microscopic galvanic cells can be made within a metal's matrix between precipitates at the grain boundaries and the grains. If the precipitate is the anode and the grain is the cathode, the material at the grain boundary will go into solution while it is corroding and form internal cracks and eventually fail. The potential between the two metals can be measured by referring to an electromotive force table found in reference books (see Figure 7.11). The galvanic corrosion rate also can be magnified or increased by the area effect. The corrosion rate will increase when a small anode and a large cathode are coupled. An example is steel rivets in a copper plate. This corrosion reaction will occur much more rapidly than copper rivets in a steel plate.

Galvanic cells can also be used to protect materials. Cathodic protection is a

Figure 7.11
*Electromotive force table. More noble metals
are cathodic, and more reactive or corrosive
metals are anodic (adapted from Reference 4).*

Figure 7.12
*Galvanic series of
alloys in seawater
(adapted from Reference 4)*

Metal-Metal Ion Equilibrium (Unit Activity)	Electrode Potential vs. Normal Hydrogen Electrode at 25°C, Volts	
Au-Au^{+3}	+1.498	
Pt-Pt^{+2}	+1.2	
Pd-Pd^{+2}	+0.987	
Ag-Ag^{+}	+0.799	
Hg-Hg$_2^{+2}$	+0.788	**Noble or cathodic**
Cu-Cu^{+2}	+0.377	
H$_2$-H^{+}	0.000	
Pb-Pb^{+2}	−0.126	
Sn-Sn^{+2}	−0.136	
Ni-Ni^{+2}	−0.250	
Co-Co^{+2}	−0.277	**Active or anodic**
CD-CD^{+2}	−0.403	
Fe-Fe^{+2}	−0.440	
Cr-Cr^{+3}	−0.744	
Zn-Zn^{+2}	−0.763	
Al-Al^{+3}	−1.662	
Mg-Mg^{+2}	−2.363	
Na-Na^{+}	−2.714	
K-K^{+}	−2.925	

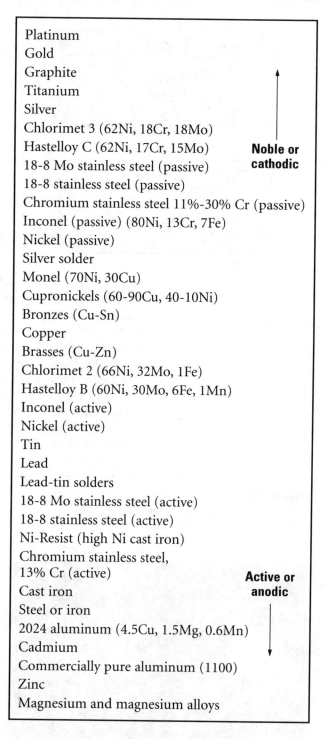

Platinum
Gold
Graphite
Titanium
Silver
Chlorimet 3 (62Ni, 18Cr, 18Mo)
Hastelloy C (62Ni, 17Cr, 15Mo)
18-8 Mo stainless steel (passive)
18-8 stainless steel (passive)
Chromium stainless steel 11%-30% Cr (passive)
Inconel (passive) (80Ni, 13Cr, 7Fe)
Nickel (passive)
Silver solder
Monel (70Ni, 30Cu)
Cupronickels (60-90Cu, 40-10Ni)
Bronzes (Cu-Sn)
Copper
Brasses (Cu-Zn)
Chlorimet 2 (66Ni, 32Mo, 1Fe)
Hastelloy B (60Ni, 30Mo, 6Fe, 1Mn)
Inconel (active)
Nickel (active)
Tin
Lead
Lead-tin solders
18-8 Mo stainless steel (active)
18-8 stainless steel (active)
Ni-Resist (high Ni cast iron)
Chromium stainless steel, 13% Cr (active)
Cast iron
Steel or iron
2024 aluminum (4.5Cu, 1.5Mg, 0.6Mn)
Cadmium
Commercially pure aluminum (1100)
Zinc
Magnesium and magnesium alloys

Noble or cathodic

Active or anodic

Figure 7.13
*Sacrificial anodic or
cathodic protection is
used in galvanized steel*

method of coupling a more anodic or corrosive metal to the material that is being protected. Here, the sacrificial anode will corrode and protect the needed material. This system is used for galvanized steel where the zinc that is plating the steel is more corrosive than the base metal. Sacrificial anodes are also used in underground oil tanks and boat hulls.

Selective leaching, or parting corrosion, occurs when one element of an alloy is removed from the solid solution. One example is with brasses and is called dezincification. The zinc leaves the unit cell and goes into solution. The copper that is left is very brittle and porous from the holes left behind from the missing zinc.

Stress corrosion occurs when corrosion and mechanical stress combine to produce a failure. The introduction of corrosion reduces the stress needed for the material to fail. Stress corrosion cracking is the production of cracks in a material with the introduction of corrosion and stress. Here, the stress may also be applied by a difference in the coefficient of thermal expansion between materials. The coefficient of thermal expansion will be discussed in the Stress-Strain section. An example of stress corrosion is sensitization of stainless steels when welding. The heat-affected zone (HAZ) of the weld promotes the precipitation of $M_{23}C_6$ carbides. More specifically, these carbides are $Cr_{23}C_6$. Stainless steel is corrosion-resistant when there is over 11% chromium (Cr) present in the matrix. This Cr forms a protective oxide on the surface that resists corrosion. When the $Cr_{23}C_6$ forms in the HAZ, the area adjacent to the carbides becomes lean in Cr. The Cr-depleted area (<11%) corrodes like carbon steel and eventually fails. The weld is not the failure area but the area adjacent to the weld, the HAZ. The solution to sensitization is to use a low-carbon stainless steel (316L, 304L, etc.) so that carbide precipitation is not promoted.

The addition of ions (Cl^-, Br^-) also promotes the chemical reaction of corrosion. In the case of Cl ions reacting with aluminum, the following reaction

Figure 7.14
*The effects of Cl-rich paint remover
on an Al pad with Au ball bonds*

Figure 7.15
Flux residue causing solder joint corrosion

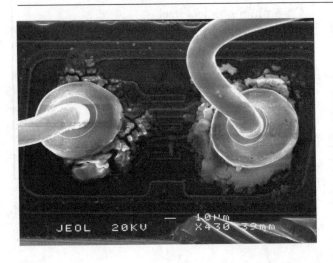

occurs, resulting in aluminum chloride and hydrogen bubbles offgassing:

$$-6HCl + 2Al \rightarrow 2AlCl_3 + 3H_2 \qquad (7.8)$$

Ionic contamination is also left with the more aggressive flux residues and volatile organic acids. This will promote corrosion of solder joints and other component degradation.

Dendritic growth is the movement of material across a migration path that has similar requirements to basic corrosion with the addition of a DC electrical field. Silver (Ag) migration is one of the most common dendritic growth processes in modern electronics. It can cause shorts in circuits that may be intermittent, but most important, the migration can only be seen *in situ*. A reliability-screening test would have to be put into place to simulate the life situations to propagate Ag migration. The requirements for Ag migration are:

Figure 7.16
Ag migration process

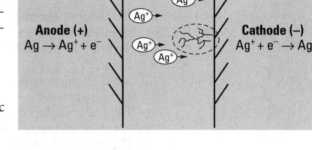

- moisture,
- DC electric field across the migration path,
- Ag epoxy on an anodic site relative to a cathodic site, and
- contamination.

It should also be noted that Ag migration has been observed without an electric field in the presence of high levels of F^- and Cl^-, although this is temperature-dependent.

Figure 7.17
Ag dendrites forming after 1000 hours of 85°C/85% RH and 2.5 V of DC bias of Ag epoxy attach. The Ag crossed a spacing of 50–125 microns.

Figure 7.18
EDS x-ray map of Ag migration on the wire bond of a diode. Ag is a plating layer in the mesa of the diode.

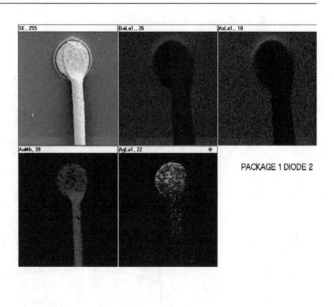

PACKAGE 1 DIODE 2

7.2.5 Stress-Strain

The stress-strain relationship is a direct result of mechanical properties of materials. These properties may be hardness, toughness, and the ductility or brittleness of a material. Mechanical failures have a large dependence upon the physical properties. The forces upon the material also will determine if the failure is pure tensile-, shear-, fatigue-, or creep-related. Some definitions pertaining to stress-strain are as follows:

$$EngineeringStress = \sigma = \frac{load}{original.area} \qquad (7.9)$$

$$EngineeringStrain = \varepsilon = \frac{change.in.length}{original.length} \qquad (7.10)$$

- Modulus of elasticity (psi or MPa) – stress-strain in elastic range (slope of stress-strain curve)
- Tensile strength or UTS (psi or MPa) – maximum stress on the stress-strain curve
- Yield strength (psi or MPa) – stress at which 0.2% permanent or plastic strain is present
- Fatigue or endurance limit – stress required to produce failure in a material subjected to a specific number of cycles (loading and unloading)
- Creep – plastic deformation over time under a constant stress
- Toughness – the combination of hardness and ductility (the area under the stress-strain curve)

Common properties of materials can be obtained in reference books. Please note that properties such as modulus of elasticity, UTS, and creep data will change with temperature, strain rate, load, and amplitude of cycles. Data from one source must be carefully compared to data from a second source.

The coefficient of thermal expansion is an important factor when mating two material systems together in an environment that will experience temperature change. Materials will expand or contract under temperature changes. In these cases, a stress is created by differences in CTE:

$$\sigma = E(\partial_1 - \partial_2)\Delta T \qquad (7.11)$$

Figure 7.19

Stress-strain curves for various alloys. Note the decrease in stress needed to complete the fracture after the sample reaches its UTS. This is due to the necking of the specimen of ductile materials (from Reference 4, with permission).

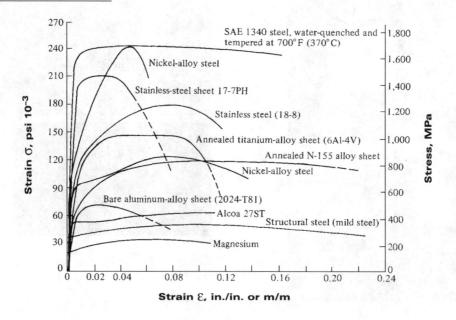

Moisture Vaporization During Heating

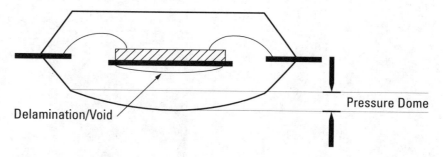

Delamination/Void

Pressure Dome

Plastic Stress Fracture

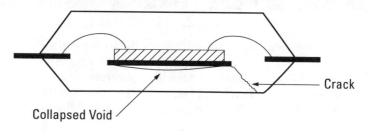

Crack

Collapsed Void

---

where

σ = the stress,

∂_1 = the coefficient of thermal expansion for material 1,

∂_2 = the coefficient of thermal expansion for material 2, and

E = the modulus of elasticity.

Thermal fatigue is multiple thermal cycles that induce material degradation and eventually failure. Thermal shock is one thermal cycle that induces failure. CTE data can be acquired in *The ASM Handbooks*.

Popcorn cracking is a phenomenon that combines moisture absorption and thermal expansion to create shear stresses at interfaces in plastic IC packages.

Figure 7.20 (above)
Schematic of popcorn cracking

Figure 7.21
Scanning acoustic image of delamination at the plastic encapsulant to the paddle interface after popcorning. The first eight were run through a reflow oven, and the last two are virgin.

Figure 7.22
Cross-section of popcorning showing the delamination between the leadframe and the plastic encapsulant

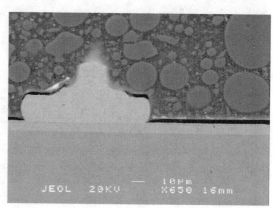

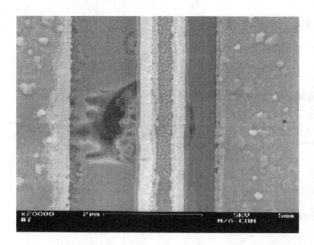 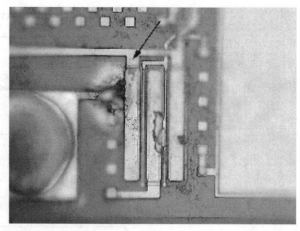

Figure 7.23
Optical and SEM images of ESD strikes

Moisture ingresses through the plastic to the surface of the package leadframe. Upon heating of the package for soldering processes, the moisture then vaporizes and causes internal pressure and delamination. The void then collapses and forms cracks in the plastic at stress concentration areas. Other effects may occur such as pulled bond wires and cracked die.

7.2.6 Electrostatic Discharge (ESD)

ESD is the rapid transfer of electrons from one potential to another. This results in hard failures and latent product defects. A latent defect is one that is not seen in final testing of the product but results in a component being in a weakened state. The customer then puts the part in use, and eventually the weakened component fails under a less-than-expected load or in a fraction of the designed life. ESD has been suspected to be the largest root cause for failure in many major microelectronic companies.

Failure modes after an ESD event can include anomalies in dielectrics such as gate oxides, insulation, and nodes bridging dielectrics. In junctions, metal spiking is common. ESD simulators are available to help determine the threshold that a specific component can tolerate. It is also a complementary tool for the failure analyst to repeat a specific event that happened in the field. Electrical Overstress (EOS) is a phenomenon where a component is subjected to excess current, resulting in failure. These events can be caused by loss of gate control, test errors, and system spikes. EOS is usually more destructive than ESD and is easier to find and evaluate.

Figure 7.24
FE-SEM images of nodes bridging a dielectric and an FIB section of an ESD event

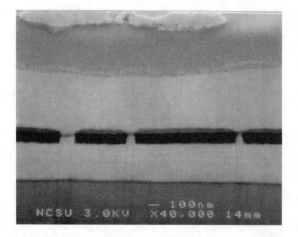 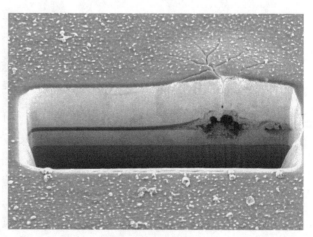

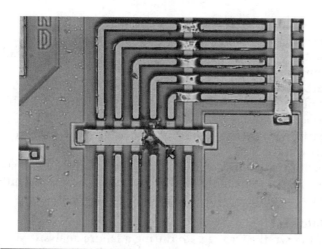

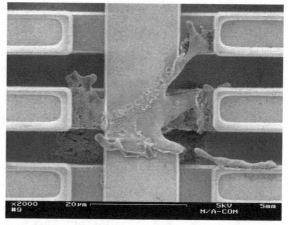

7.3 Analysis Flow

Section 7.3 will conform to the following flow chart for an analysis procedure.

 1. Define Problem

 2. Collect Data

 3. Define Analysis

 4. Execute Plan

 5. Identify Root Cause

 6. Take Preventive Action

 7. Document Database

 8. Celebrate

7.3.1 Define the Problem – Asking the Right Questions

The key to a good analysis is to ask the right questions:

1. Why is the component or part considered not operable?
2. What environment was the component or part in when it ended up in its present state?
3. What is the history of the component or part? Has the item undergone several stressful environments before failure occurred? Did the final environment prompt the failure?
4. How was the component or part taken out of service?
5. Is the obvious fracture/failure the cause for the failure or is it a result of the failure?
6. How many components/parts under identical conditions resulted in the same failure?
7. What materials are the components/parts made of? Are they the correct materials?
8. Why is an analysis needed?

Correct questions result in more relevant answers. The proper questions have to be asked to drive a successful analysis. Usually, both the failure mode and failure mechanism (e.g., root cause for the failure) have to be identified.

Figure 7.25 (above)
OM and SEM images of an EOS event. Note the high degree of destruction compared to an ESD event.

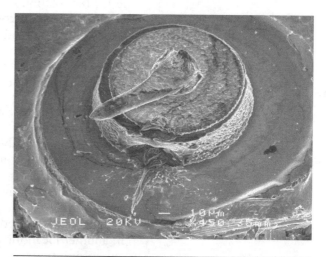

Figure 7.26 (above)
SEM image of the EOS of a diode mesa

Here are a few examples of the differences:

1a. The component/part does not operate properly because of shorted leads.
1b. The root cause of the failure is ionic metal migration between leads because of contamination, thus causing a short between the leads.
2a. The component/part does not operate properly because of a blown trace.
2b. The root cause of failure is electrostatic discharge (ESD) during assembly, which caused the trace to blow.
3a. The component/part does not operate properly because of a fracture in the solder joint between the lead and the board.
3b. The root cause of failure is the brittle tin/gold (Sn/Au) intermetallic phase that formed in the solder joint because the leads were not sufficiently wicked with Sn prior to soldering, thus causing the solder to fracture.

Develop a Problem Statement

It is quite easy to drift down the wrong path during a failure analysis. Many questions arise during the analysis that may or may not be relevant to the problem at hand. Many times, more questions can be created by the analysis than are answered by the procedures. This is why a problem statement should be developed before any plan of analysis is made or any action is taken. This **problem statement** should keep the analyst focused on the problem. After each action taken, the analyst should revisit the problem statement and determine whether the solution is closer or farther away. The analyst should then tailor the analysis and proceed down the correct path. The problem statement should be carefully prepared to address the real issues and not the superficial problems. In the best-case scenario, all of the parties involved with the component/part develop the problem statement together. This helps the analyst greatly by reducing redundant or unnecessary work and ultimately saving valuable time.

7.3.2 Data Collection – Product History

After the problem statement has been developed, the next step is to find all the information possible on the component/part up to the time of failure.

- Where was the part made?
- What type of environment is the manufacturing process in? Is it a clean room, static-sensitive area, temperature-controlled area, humidity-controlled area?
- Was the part received directly from manufacturing or was it post-processed in another location?
- What condition was the part in when it was received? What was it packaged in? Was it static-sensitive?
- Were incoming tests performed on the part to verify the failure? Are there travelers that document tests to show the part was working after manufacturing?
- At the time of failure, what was the environment (temperature, humidity, time at extreme temperatures)?
- Were there other external or internal stresses on the part during operation?
- Were the operating stresses in accordance with the part specifications?
- Were screen tests performed on the part prior to shipping?
- What is the failure mode? How did the device fail?
- Where did the failure occur?
- What was the last assembly step before the failure was discovered?
- Were there any deviations from normal that occurred at or near the time of failure?
- Was the device being used in a new application or altered conditions?
- Is the failure isolated to one particular customer or application?

- How was the failure isolated to this device?
- Has the failure occurred previously?
- Was the failure intermittent?
- Did replacing the device fix the problem?
- What were the conditions at the time of failure at customer, supplier, or end-user site?
    1. Operational electrical stress applied to device by circuit
    2. Intentional external stress applied to device (burn-in)
    3. Environmental stress (temperature, humidity, location, etc.)
- How was the device handled before submittal for analysis?
    1. Effects of removal of device from the place of operation
    2. Packaging
    3. Environmental effects
- Device-related data: part number, date code, lot information, schematic (device and surrounding circuitry)
- Does an FMEA exist on the device or process?

While this is a long list of questions, the more answers acquired, the better the analysis. Another quality resource is the reliability laboratory of the manufacturer of the part. In some cases, the reliability laboratory will have a database of past analyses about this type of failure.

7.3.3 Define Analysis – Establish a Failure Analysis Plan

Once the key questions are answered, a failure analysis plan can be made with the appropriate tests. Depending upon the type of failure, the analytical tools employed will vary. Electrical opens or shorts will need a curve tracer, a probe station, or a Scanning Electron Microscope (SEM). Material failures, such as delaminations, metal migration, or contamination, may need chemical analyses such as Energy Dispersive X-Ray Spectroscopy (EDS), Scanning Auger Microanalysis (SAM), or Fourier Transform Infrared (FTIR). Determining which of these to use will be discussed later in the chapter.

Plan all tests in order of severity. Nondestructive tests should be performed first so that evidence is not destroyed before it can be collected. Plan tests so that one test does not interfere with subsequent tests (i.e., perform a leak test before the package is opened).

7.3.4 Execute a Failure Analysis Plan

Some basic guidelines to execute a concise, systematic failure analysis plan follow:

- *Perform analysis procedures determined in the analysis plan.* The analysis plan structures analyses in a sequence that does not destroy any evidence (e.g., SIMS analysis removes material from the surface).
- *Document the condition of a device before and after each test.* It is imperative to photodocument the part after tests that alter the part. By doing this, there will always be documentation as to what is real evidence and what is an artifact of analysis.
- *Document the results of all tests.* All results should be documented in hard copy, even if there is no result. No result from a test can be as informative as achieving results.
- *Review the data collected at each step in the analysis plan.* The analysis plan can be modified to account for the results. If the data show that the plan is not on track, stop and revise the plan. In some cases, a meeting with the involved parties may be useful to determine the next steps.

• *Performed controlled experiments.* In some cases, the data collected from the analysis provide no clear indication of the failure mode. Controlled experiments may provide more indication of the root cause of failure. The number of variables should be kept to a minimum so that the experiments are useful.

7.3.5 Determining the Failure Mechanism

Determine the *root cause* or *failure mechanism* based on the results of the analyses and related research. Give the root cause based on the data and not on a perception of what would be good to find or easy to fix. Some parties may be unwilling to accept the conclusion that their product or process caused the failure, but failure analysts must be impartial and report their findings.

7.3.6 Determining the Corrective Action

Determine and implement a corrective action sufficient to preclude the failure from recurring. This can require cooperation of all parties, from manufacturing to sales. Common effective corrective actions usually are:
• Writing or revising procedures or documents to improve the process and product.
• Changing design or materials.
• Documenting and controlling training.

7.3.7 Document/Database

Document and use a database for all information acquired during the analysis. As stated previously, no results are sometimes as informative as reams of data. The failure analysis report should include the following:
• Problem statement,
• Part or component history,
• Failure analysis procedures,
• Findings of all tests and inspections,
• Conclusion identifying the root cause of the failure, and
• Corrective action statement outlining the actions taken to prevent recurrence.

The report should tell the story from the beginning to the end. Include all of the part's history along with all of the data. The reports should then be put into a database or filed to access the data easily should a similar failure occur.

7.4 Failure Analysis Example

Figure 7.27
Failed attenuator
(see Reference 6)

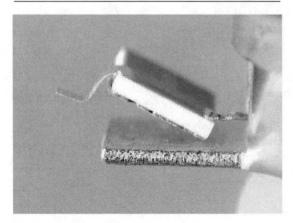

Problem: An attenuator "popped" off or delaminated during a solder-reflow process. The information provided by the engineer stated the materials involved and the temperatures used to reflow the solder.

Problem statement: Why did the attenuator pop, and how do we prevent it in the future?

Analysis: The first step in the analysis was to photodocument the failure using optical microscopy. The next step was to determine the layers in the platings that delaminated. Here, Scanning Electron Microscopy (SEM) and Energy Dispersive X-Ray Spectroscopy (EDS) were utilized.

Through literature and phase diagrams (Figures 7.28 and 7.29), it can be seen that over 4% Au in tin/lead (Sn/Pb) solder can cause an embrittled joint, called gold embrittlement. The results of the EDS show that there is gold (Au) in the solder joint (Figure 7.30). The next step in the failure analysis process was to take a cross-section of a virgin part that had seen no reflow temperatures, using standard metallurgical techniques to determine whether any Au is in the joint.

Figure 7.31 is an SEM micrograph using Backscattered Electron Imaging to show the $AuSn_4$ intermetallics in the solder joint. The digital x-ray map using EDS shows the distribution of the elements throughout the joint: It can be seen that there is Au through the entire joint. There is enough evidence to show Au embrittlement and to answer the question of why the failures are occurring.

The next step is to answer the question in the problem statement of how to prevent this in the future. Most actions include consultation with the manufacturing engineer to ensure feasibility. In this case, it is a metallurgical problem, and the only solution is to remove the Au from the joint. This either entails putting on thinner Au so the percentage never exceeds 4% in the joint or tinning and wicking the Au off of the part before the soldering process.

Document and Database: **All** of the previous information and data acquired should be copied and filed in the database to use for future reference, to avoid performing the same analysis twice.

Figure 7.28

Graph showing mechanical degradation with increasing Au content (see Reference 3)

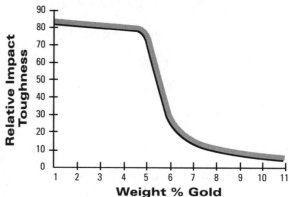

Figure 7.29

Phase diagram of Au/SnPb (see Reference 3)

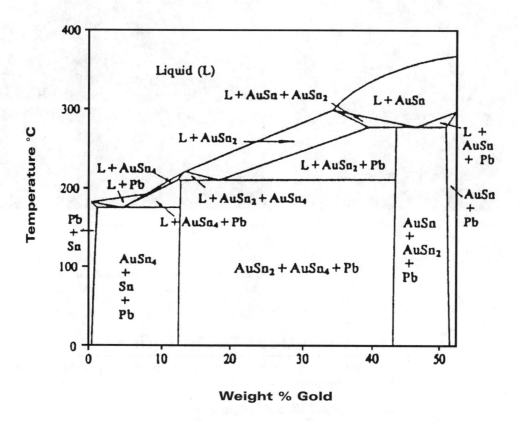

Weight % Gold

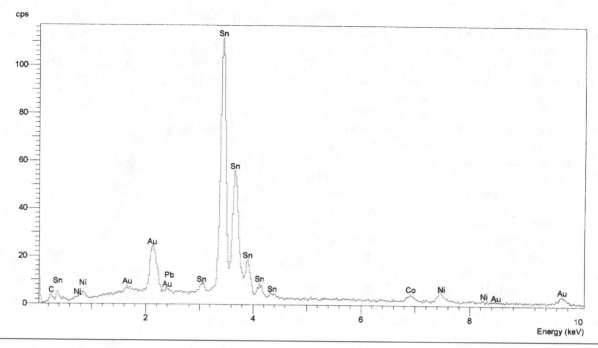

Figure 7.30 (above)
EDS spectrum showing gold (Au) present in solder

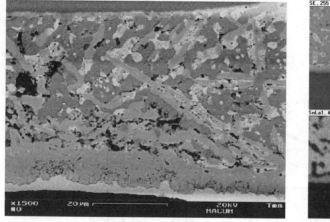

Figure 7.31
Backscattered electron imaging showing intermetallics

Figure 7.32
EDS x-ray map of a solder joint. Note the Au/Sn layer between the Cu base material and the solder. This layer is $AuSn_4$ and will cause gold embrittlement.

7.5 Analytical Techniques

The following is a list with descriptions of analytical techniques that are often used for analyses.

7.5.1 External Visual Inspection

External visual inspection is nondestructive investigation of parts or components using visual eye or optical light microscopy (OM). In addition to

providing a photodocument before each analysis step, OM can be used to find:
- package surface contamination which can cause electrical leakage,
- fractured or broken dielectrics or glass seals,
- fractures or gaps in weld seams,
- broken leads,
- discoloration from burns and overheating, and
- correct part numbers.

Figure 7.33
OM micrograph showing stain on SMT device

7.5.2 X-Ray Radiography

Principle of Operation

X-ray radiography uses invisible, highly penetrating, short wavelength electromagnetic radiation to nondestructively obtain an image of details that cannot be seen with OM inspection because of hidden details. Areas where more radiation is transmitted (light elements, thin sections) produce a higher signal. Two techniques are most commonly employed:
- Conventional radiography makes use of the transmitted radiation to expose film that can be conventionally developed. This is the most commonly used technique and is seen extensively in the medical industry.
- Real-time x-ray methods involve the projection of an image as the part is being irradiated, eliminating the waiting time to develop the film. This technique also adds to the flexibility, where the part can be rotated and tilted in real time.

Applications
- Broken bonds;
- Lifted bonds (side view);
- Misplaced bonds;
- Crossed wires or wires shorted to the die, leadframe, or case;
- Die attach voiding (epoxy, solder, etc.); and
- Misalignment of internal components.

7.5.3 C-Mode Scanning Acoustic Microscopy (C-SAM)

The C-SAM is an analytical device that uses acoustic reflection to reveal voids, cracks, disbonds, and delaminations in the bulk of samples. This is a nondestructive test as long as the part can be submersed in deionized water.

Figure 7.34
X-ray image showing die and bond wires in a plastic-encapsulated IC

Principle of Operation

The C-SAM mechanically rasters a piezoelectric transducer in a pattern over a sample to produce an image. A focused spot of ultrasound, produced by an acoustic lens, is brought to the sample by a coupling medium (DI water). An acoustic pulse enters the sample and echoes back at specific interfaces within the bulk of the sample. The return time of the pulse determines the image.

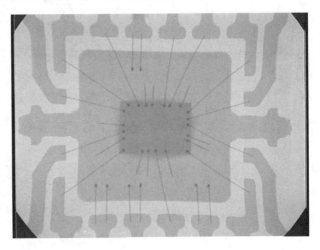

Applications
- Identifies voids, cracks, and disbands;
- Identifies die or substrate attach delamination;
- Identifies adhesion failures that are not detectable by x-ray analysis;

Figure 7.35 (below)
C-SAM image showing delamination between the plastic encapsulant and the die paddle and die

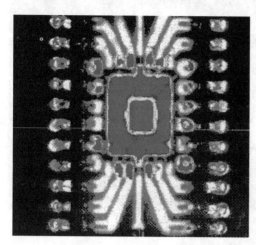

- Complements data acquired by x-ray examination; and
- Allows imaging through dense metals that x-rays cannot penetrate.

Quality Control – Comparison of good and bad samples, percent bonding coverage, and bonding integrity.

7.5.4 Scanning Electron Microscopy (SEM)

SEM is a technique used to image topographic or microstructural features on polished or rough surfaces at ultra-high resolution (as high as 10 Angstroms) and still achieve good depth of field. SEM is usually a nondestructive technique (some samples exhibit severe electron beam damage) if a conductive coating is not needed.

Principle of Operation

A current is put through a filament to emit a primary electron beam down the column of the instrument to penetrate into the sample. The image is generated by rastering this primary electron beam over an area and synchronously displaying on a cathode ray tube the spatially magnified secondary (or backscattered) output. This results in an image in which bright and dark areas correspond to areas of high and low electron output. Secondary electrons (SEI) escape the sample surface from approximately 300 Angstroms. The primary beam penetrates approximately 1–3 µm into the sample, thus producing a high-resolution image of the topography. Presently, there are several types of SEMs: conventional, field-emission (FE), and environmental types.

A conventional SEM uses either a tungsten or a lanthanum hexaboride (LaB_6) filament with a gun vacuum of 10^{-7} Torr. Depending upon sample interaction with the beam, the usable resolution limit is approximately 50 Angstroms ($10\times$ to $100,000\times$). The usable accelerating voltage range is 2 kV to 40 kV, with the resolution falling off greatly at the lower voltages. The advantages of using a LaB_6 filament is a brighter image and longer life. A conventional SEM needs to have a conductive sample, either naturally or by coating.

Figure 7.36 (below)
Schematic showing electron beam interaction (from Reference 8, with permission)

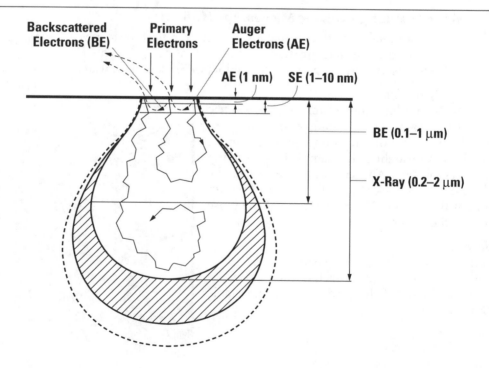

An FE-SEM uses a tungsten filament whose diameter is an order of magnitude finer than a conventional SEM and with an ultrahigh vacuum of 10^{-10} Torr. The resolution capability is much greater than a conventional SEM at approximately 10 Angstroms (10× to 800,000×) even at lower accelerating voltages. The voltage range is from 0.1 kV to 30 kV, with optimum resolution at the lower voltages. The use of low voltages allows nonconductive samples to be imaged.

An environmental SEM uses variable column pressure to produce a low vacuum in the sample chamber to accommodate wet and nonconductive samples. Environmental SEMs can be tungsten or LaB$_6$. Hot and cold stages can also be used to watch changes in structure or phases that occur in materials.

Figure 7.37
SEI SEM image of an airbridge at a 45° angle showing the depth-of-field capability of an SEM

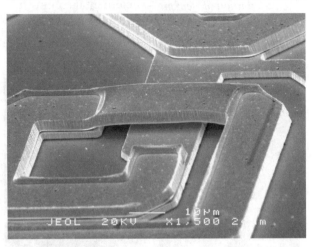

SEM Imaging Modes
Secondary Electron Imaging (SEI) Mode
- Low-energy electrons are collected from surface by a high-voltage-biased collector around the SEI detector.
- High electron yield produces bright image with low noise.
- Can collect electrons from around corners.
- Does not provide very much topographical information because of beam penetration.
- Good depth of field.
- High resolution.

Backscattered Electron Imaging (BEI) Mode
- Higher energy electrons backscattered from within the sample, away from the principal beam interaction.
- Low electron yield requires higher beam current to produce low-noise image.
- Electrons collected are straight line from sample to detector (no collector).
- No collection from around corners; hence, good topographical information.
- Lower resolution than SEI due to electron scattering and larger primary beam.

Electron Beam-Induced Current (EBIC)
- Views the formation of junction/hole pairs in a semiconductor material.
- Uses the electron beam as the excitation source (does not require external bias).
- High-gain amplifier is connected to two device leads and synchronized with the CRT trace.
- Conductor traces are not visible.
- Passivated devices can be viewed successfully.

Voltage Contrast (VC)
- Electrical bias is supplied to the device through interconnects in the SEM chamber.
- Varying voltage levels on the device change the secondary electron emission, producing a visible contrast difference.

Applications
- Material evaluation and characterization – determine surface morphology and visual inspection;
- Fracture surface analysis (fractography);
- Microstructural characterization in metals, ceramics, and geological samples;
- Failure analysis – distribution of contaminants (BEI), failure mode, failure mechanisms, and reverse engineering;

• Inspection and characterization of defects in integrated circuits; and
• Quality control – comparison of good and bad samples, plating thickness measurements, and process evaluation.

7.5.5 Energy Dispersive X-Ray Spectroscopy (EDS)

EDS is a bulk technique that qualitatively and quantitatively identifies the elemental composition of materials analyzed in an SEM. EDS has a depth penetration of approximately one micron. Depending upon the detector used, all elements as heavy or heavier than beryllium (Be) on the periodic table can be detected by EDS with a detection limit of greater than 0.1 wt %.

EDS output is in the form of an area or spot (one-micron spatial resolution) spectrum, an x-ray map, or line scan that shows the distribution of elements over an area.

Figure 7.38

Digital x-ray map showing distribution of elements across an area

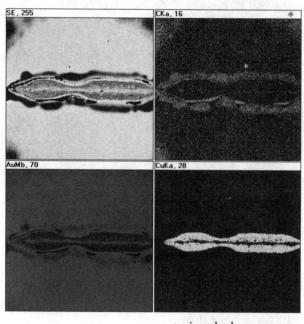

Principle of Operation

The electron beam from an SEM is used to scan across the surface of a sample. X-rays are generated from the atoms on the top microns of the sample's surface. The x-rays are produced when outer shell electrons drop down to inner shell vacancies resulting in secondary electron production. The energies of the x-rays are defined by the energy differences between the outer and inner shell electron energy levels. Each element is characterized by the x-ray energy detected.

The EDS system collects the x-rays, sorts them by energy, and displays the number of x-rays (peak intensity) versus energy. An x-ray map showing the distribution of any specific element can be collected by rastering the beam over an area and synchronously displaying the output of x-rays of the appropriate energy.

Applications

• Materials evaluation – identification and verification of contaminants, alloys and intermetallics, material composition, and diffusion profiles;
• Identification of corrosion products; and
• Quality control – material verification, alloy identification, and plating specification certification.

7.5.6 Scanning Auger Microanalysis (SAM)

SAM is a surface analysis technique that determines qualitative and semi-quantitative elemental composition and chemistry information of surfaces and interfaces. SAM has a sampling depth of 10 to 30 Angstroms, which will provide elemental information of films as thin as a few monolayers.

SAM can also show distribution of elements in map form as well as depth distributions when ion milling is employed. SAM detects lithium (Li) and heavier elements on the Periodic Table in practical concentrations above 1.0 wt %.

Principle of Operation

An electron beam strikes a sample and ejects orbital electrons from around the atoms that are left in the high-energy or excited state. If the ejected electron is from an inner core shell, the excited atom can relax to lower energy by a process in which an electron in an outer shell falls into the vacancy in the inner shell. When this occurs, transitional radiation is generated and is dissi-

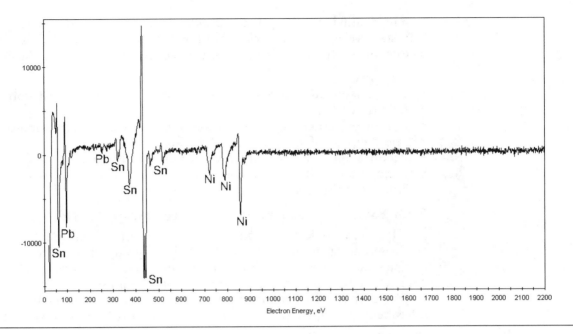

Figure 7.39
SAM spectrum of a solder-dipped lead with Ni diffused to the surface

pated, either by the ejection of another valence electron, called an Auger electron, or by the emission of an x-ray photon. The Auger electron has a kinetic energy characteristic of the atom. Detection and energy analysis of the Auger electrons lead to the identification of the target atoms.

Depth profiles are conducted when an ion beam that removes consecutive atom layers from the sample surface bombards the sample surface. Subsequent Auger analysis on the new surface allows concentration gradients to be obtained.

Applications

- Materials evaluation – identification of surface contaminants, verification of surface homogeneity, diffusion, and interface studies;
- Composition and morphologies of thin films;
- Failure analysis – corrosion and oxidation products, stain identification, and material delamination analysis;
- Identification of second phase inclusion or particulates; and
- Quality control – comparison of good and bad samples, verification of surface process modification, and relative thickness determination on thin films.

7.5.7 X-Ray Photoelectron Spectroscopy (XPS)
Electron Spectroscopy for Chemical Analysis (ESCA)

XPS is a surface analysis technique that determines qualitatively and quantitatively elemental composition and valence states and/or bonding environment of an atom near the surface of the sample. XPS accommodates solid samples that need not be conductive and can detect elements as heavy as or heavier than lithium (Li) ($Z \geq 3$) on the periodic table.

Principle of Operation

A sample is irradiated with magnesium (Mg) or aluminum (Al) source x-rays, which causes the ejection of photoelectrons from the surface. The photoelectrons ejected from the surface have an energy equal to the x-ray energy ($h\nu$), where ν is the x-ray frequency, minus the binding energy of the electron in the shell from which it was ejected. The electron binding energy, which is measured by a high-resolution electron spectrometer, can provide information in identifying the target atoms and, in many cases, determine the oxidation state, valence state, or chemical bonding environment of those atoms. The depth penetration of the beam is typically 30 Angstroms from the surface.

Applications
- Determination of valence states and/or bonding environment of atoms near the surface,
- Measure oxidation states of metal atoms in some metal oxide surface films,
- Determination of surface carbon (graphite or carbide),
- Identification of organic functional groups in polymers,
- Depth profiling of materials, and
- Surface identification of materials.

7.5.8 Fourier Transform Infrared Spectroscopy (FTIR)

FTIR is an analytical technique used to fingerprint or identify organic material, some inorganic materials, and functional groups present in a sample. Samples can be solids, liquids, solutions, or gases. The spatial resolution is approximately 50 μm with a microscope but a 3-mm sample is preferred for maximum sensitivity. Samples must also be infrared active (no metals) to produce the spectra in the range of 2 to 25 μm wavelength (5000 to 400 cm^{-1}).

Principle of Operation

A beam of infrared radiation (far-infrared wavelengths) is transmitted through a sample, and the constituents preferentially absorb certain wavelengths of radiation. The light travels into a sensitive infrared detector, and a computer performs a Fourier transform to convert the time-modulated intensity changes at the detector into an absorption spectrum vs. wavelength for the sample. Identification of the spectral absorption bands allows identification of the composition of the sample.

Figure 7.40

FTIR spectrum showing the comparison of virgin tape and IR exposed tape

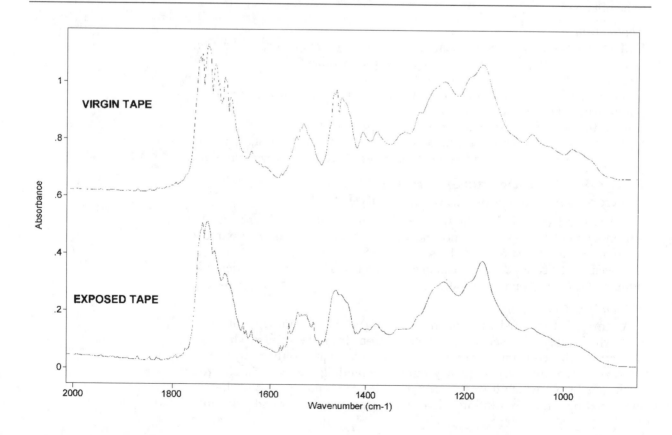

Applications
- Identification of polymer coatings, fibers, packaging material, and composites;
- Trace gas analysis;
- Identification of most solid or liquid organics and polymers;
- Failure analysis – identification of contaminants on microelectronic packages and devices, organic stains, process fluids, and component degradation or decomposition;
- Identification of adhesives, cleaners, and solvents;
- Quality control – comparison of good to bad samples;
- Verification of parts and solvent cleanliness; and
- Biomaterials.

7.5.9 Infrared Imaging

An infrared imaging system is an analytical technique in which infrared is used to produce thermal images on a sample. The surfaces to be imaged must be a black body (high-emissivity) radiator to obtain direct temperature measurements. Low-emissivity or reflective surfaces, such as gold, must be coated or undergo an emissivity correction to obtain an accurate temperature reading.

Figure 7.41
IR image of active component

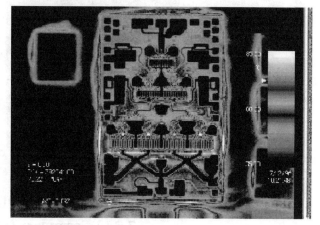

Principle of Operation

An infrared imager detects infrared radiation (IR), converts it to electronic signals, and displays real-time images that show the intensity or level of the radiation.

Applications
- View hot spots on circuitry (die, board, or system level) while powered or biased;
- View the thermal characteristics of a product under development;
- Quality control – verify operating temperatures of components to theoretical temperatures, compare thermal characteristics of a failed device with that of a known good device; and
- Failure analysis – identify the failure temperature of a biased component.

The limitations of the approach are as follows:
1. The surfaces to be imaged must be a black body (high-emissivity) radiator to obtain direct temperature measurements.
2. Reflective (low-emissivity) surfaces, such as gold, must be coated or undergo an emissivity correction to obtain accurate temperature readings.

7.5.10 Focused Ion Beam (FIB)

FIB is an analytical technique that uses an ion beam to image and sputter the surface of a sample. The sputtering technique uses the ion beam to prepare a highly defined cross-section with 0.2-μm accuracy. The clean, smear-free cross-section shows more delineated interfaces than manual grinding and polishing metallurgical techniques.

FIB can also be used for device modification, where a metal is deposited after the ion beam cut to make a connection between areas of the circuitry. This shortens time in the redesigning and processing areas.

Principle of Operation

A focused ion beam is used to sputter the surface of a sample.

Applications

• View a suspected failure site in the Z axis;
• Prepare a multilayer metallization scheme for thickness measurements; and
• Make a connection between areas of circuitry for experimental procedures.

Figure 7.42

FIB cross-section of a microelectronic circuit. Note the clear delineated layers exposed by the FIB sectioning technique.

7.5.11 Atomic Force Microscopy/ Scanning Probe Microscopy (AFM/SPM)

AFM and SPM are techniques that image surfaces with atomic or near-atomic resolution. With AFM/SPM, a Z dimension can be measured quantitatively, allowing for surface roughness measurements.

Principle of Operation

A small tip is scanned across the surface of a sample using piezoelectrically induced motions. The computer then translates these motions into a three-dimensional image of the surface. If the tip and the surface are both conducting, the structure of the surface can be detected by tunneling of electrons from the tip to the surface (Scanning Tunneling Microscopy, STM). The molecular forces exerted by the surface against the tip can probe any type of surface. The tip can be constantly in contact with the surface, it can gently tap the surface while oscillating at high frequency, or it can be scanned just minutely above the surface (see Reference 9).

Applications

• Quantitatively measure surface roughness with a nominal 5 nm lateral and 0.01 nm vertical resolution;
• Roughness of semiconductor wafers, optical components, hard disk drives; and
• Surface wear measurements.

7.5.12 Secondary Ion Mass Spectroscopy (SIMS)

SIMS is a surface-sensitive analytical technique that provides qualitative elemental and chemical analysis of the top one to five monolayers of a sample. The sputtering of the ion beam can also produce elemental depth distributions. SIMS samples must be vacuum-compatible down to 10^{-8} to 10^{-9} Torr and can cause sample damage because of the ion sputtering.

Principle of Operation

An energetic primary ion beam sputters a sample surface. Secondary ions formed in the sputtering process are mass analyzed using a double-focusing mass spectrometer to determine the atomic composition. Mass analyses may also provide chemical information from the molecular species and isotopic ratios.

SIMS provides excellent detection limits, good depth resolution, and full Periodic-Table detection; but elemental sensitivities can vary, and elemental interferences can occur. The spatial resolution of the beam is as low as 10 μm (see Reference 10).

Time-of-Flight Secondary Ion Mass Spectrometry (TOF SIMS)

A microfocused pulsed primary ion beam sputters the top surface of the sample. The ions are dispersed in time, in a time-of-flight mass spectrometer according to their velocity. The TOF SIMS can detect secondary ions over a larger mass range than static SIMS. The spatial resolution is better than 1 μm.

Applications
- Dopant profiles in semiconductor devices;
- High mass range, resolution, and mass accuracy determinations; and
- Identification of corrosion products.

7.5.13 X-Ray Fluorescence Spectroscopy (XRF)

XRF is a bulk, qualitative and quantitative characterization technique for the rapid, simultaneous, and nondestructive detection of all elements with an atomic number greater than 9.

Principle of Operation

A primary x-ray beam penetrates a sample to a depth of 10 to 100 μm and causes ejection of inner shell electrons from the atoms in the region. The electrons from the outer shells relax and drop down to fill the vacancies left by the inner ejected electrons. This produces secondary x-rays, which can be fingerprinted by the difference in energy between the outer electron that dropped down and the ejected electron. The secondary x-rays are detected by an energy-dispersive spectrometer or a crystal wavelength-dispersive spectrometer and produce a spectrum of intensity versus energy or wavelength (see Reference 11).

Applications
- Quantitative analysis of bulk elemental composition in glasses, alloys, and ceramics;
- Detection of metallic contamination on and in plastics and polymers;
- Quantitative thin film thickness and composition measurements; and
- Trace analysis to PPM levels.

7.5.14 Ion Chromatography (IC)

IC is used for qualitative and quantitative analyses of inorganic and organic anions and specific cations in aqueous solutions. The detection limits for anions is PPM or PPB under ideal conditions. Cation detection is limited to alkali and alkaline earths, ammonia, and low-molecular-weight amines.

Principle of Operation

A sample solution is injected into an eluent stream and passed through an ion exchange column. The stream is then passed through a second column, which removes the eluent ions. A conductivity meter detects the unknown ions, and the signal is proportional to concentration.

Applications
- Plating bath solution analysis;
- PCB cleanliness evaluation;
- Determination of anion contamination; and
- Solution analysis such as brines, waters, and condensates.

7.5.15 LECO (Laboratory Equipment Company) Elemental Analysis

LECO is used to quantitatively determine the amount of carbon (C), sulfur (S), oxygen (O), nitrogen (N), and hydrogen (H) in metals and alloy systems. This is a necessary analysis to complement Inductively Coupled Plasma (ICP) to determine the low atomic element compositions.

Principle of Operation

The sample analyzed is subjected to high heat or combustion to free the element under analysis in the gas form. The gases are quantified by thermal conductivity (O – 1 PPM, N – 1 PPM), measuring gas volume (H), or infrared (C – 0.001%, S – 0.005%) (see Reference 9).

Applications
- Verification of alloy designation,
- Determination of H for hydrogen embrittlement, and
- Identification of interstitial O and N in alloys.

7.5.16 Gas Chromatography/Mass Spectroscopy (GC/MS)

GC/MS provides a tool for identifying or confirming the identity of organic compounds in a variety of matrices.

Principle of Operation
GC separates a mixture into its individual components by means of a rapid scanning mass spectrometer. The mass spectrometer detects the components as they emerge from the end of the GC column. The molecules are subjected to a stream of high-energy electrons, which first ionizes them and then separates them according to their mass. These charged ions are then counted, and their mass is plotted versus intensity; this is called a mass spectrum.

Applications
- Determination of molecular weight,
- Outgassing production hardware to meet cleanliness specifications,
- Identification of internal atmospheres and contaminants in hermetic packages,
- Identification of unknown organic compounds,
- Determination of the cure and aging of polymers, and
- Identification of polymeric additives (plasticizers, flame-retardants).

7.5.17 High-Performance Liquid Chromatography (HPLC)

HPLC is a form of liquid chromatography to separate compounds that are dissolved in solution. HPLC will qualitatively and quantitatively analyze organic mixtures, and will analyze organic and inorganic compounds for impurities.

Principle of Operation
Different components in a mixture pass through the column at different rates depending upon how they interact with the carrier solvent (mobile phase) and the column packing material (stationary phase). Materials that associate with the carrier solvent pass through the column quickly, and those that associate with the stationary phase move more slowly. A material is identified by comparing the experimental retention time (time it takes to pass through the column) to a known retention time of a standard (see Reference 12).

Applications
- Identify purity of organic materials,
- Analyze for organic contaminants in solvents, and
- Monitor the stability of polymers during aging.

7.5.18 X-Ray Diffraction (XRD)

XRD is an analytical technique that will identify phases or compounds in unknown samples, determine crystal structure and lattice parameters, and determine crystal orientation. Samples identified using XRD must be crystalline, and in most cases, the spectrum can be searched in the spectra library (JCPDS cards) through a computer.

Principle of Operation
A beam of x-rays of a known wavelength is directed toward a sample. The

beam is diffracted from crystals in the lattice with peaks occurring in accordance with Bragg's Law ($\lambda = 2d\sin\theta$). Using Bragg's equation, the lattice *d*-spacings can be calculated. This *d* spacing is the fingerprint for the material and can be compared to *d*-spacings of known materials.

Applications
• Determine crystal orientation,
• Measure crystal defect density and residual stresses, and
• Determine phases present in metals and ceramics.

7.5.19 Inductively Coupled Plasma (ICP)

ICP is an analytical technique that can quantitatively and qualitatively analyze elements down to PPM levels. ICP is exceptional for identifying metal alloys and glasses.

Principle of Operation

A high-temperature plasma is produced by inductively coupling RF power into a stream of argon gas. The sample is dissolved into the plasma, and the elements emit their characteristic radiations.

Applications
• Determine metal alloy and glass designations,
• Detect trace impurities, and
• Water analysis.

References

1. ASM Handbook Vol. 3, *Alloy Phase Diagrams*, ASM International 1992.

2. *Electronic Materials Handbook, Vol. 01: Packaging*, ASM International, Materials Park, OH, 1989.

3. Bester, M. H., "Metallurgical Aspects of Soldering Gold and Gold Plating," Proceedings of INTERNEPCON, 1968, Oct. 8-10, Brighton, England, pp. 211-231.

4. Fontana, Mars G., *Corrosion Engineering*, McGraw Hill, Inc., 1986.

5. Hertzberg, Richard W., *Deformation and Fracture Mechanics of Engineering Materials*, John Wiley & Sons, Inc., 1983.

6. Bunis, C. B., and St. Armand, D., 1st Annual M/A-COM Manufacturing Conference, June 1997.

7. Bunis, C. B., "Gold Intermetallics in Solder Joints," IEEE Symposium, Oct. 1997.

8. ASM Handbook Vol. 9, *Metallography and Microstructures*, ASM International.

9. Borders, J. A., Echelmeyer, K. H., and Weissman, S. H., Short Course Program, Materials Research Society, Sandia National Labs, Albuquerque, NM, 1986.

10. Charles Evans & Associates, www.cea.com.

11. XRF, www.wco.com.

12. SCIMEDIA, www.scimedia.com.

TOPICS IN RELIABILITY

CHAPTER 8

Reliability Statistics Simplified

8.1 Introduction

Although reliability statistics is not a very easy subject, it can be simplified. A certain set of basic statistical materials is commonly used in industry. This chapter is intended to cover this set of reliability statistics as it relates to the topics in this book, including some advanced materials. The key topics discussed here are:
· basic commercial reliability statistics and concepts,
· reliability testing and statistical confidence (catastrophic and parametric),
· demonstration versus confidence tests, and
· influence of acceleration factors on test planning.

8.2 Definitions and Reliability Mathematics

The term reliability has numerous meanings. In the qualitative sense, products will either perform a required function under stated conditions for a stated period of time, or they will fail. In this sense, reliability is a probability for survival in this

Table 8.1
Selected reliability statistical functions

Measures	Equivalence	Definitions
$F(t)$	$= 1 - R(t)$	**The cumulative probability distribution function (CDF):** Probability of a component failing at time t. Alternately, probability of first failure at or before time t. Experimentally, the cumulative percent failure at each observed failure time when plotted versus time (usually on a cumulative probability paper) graphically displays this function.
$R(t)$	$= 1 - F(t)$	**Reliability function:** Probability of a component surviving a time t. Alternately, the number of units surviving at time t divided by the initial number of units.
$f(t)$	$= \dfrac{dF(t)}{dt}$	**Probability density function (PDF):** Probability of failure at an instant (a time period that is infinitesimally small). Experimentally, it is the instantaneous slope at time t found on the CDF plot.
$\lambda_{Cum}(t)$	$= F(t)/t$	**Cumulative failure rate:** Cumulative failure rate of a component at time t. Experimentally, this is cumulative percent failure at time t divided by the observed failure time t for each observed failure point when plotted versus time (usually on log-log paper) graphically displays this function. A linear relationship can exist to the hazard rate (see Appendix A).
$\lambda(t)$	$= \dfrac{f(t)}{R(t)} = -\dfrac{1}{R(t)}\dfrac{dR(t)}{dt}$	**Instantaneous failure rate, hazard rate, or just the failure rate:** Probability of failure in unit time of a device that is still working. The instantaneous rate of failure for devices of a population that have survived to time t.
MTBF & MTTF	$= \dfrac{1}{\text{Constant Failure Rate}}$ $= \dfrac{1}{\lambda}$	**Mean Time Between Failure (MTBF), Mean Time To Failure (MTTF):** Expected length of time a system/unit will be operational. MTBF is the preferred term instead of MTTF when repairs are involved. Both are the inverse of the failure rate when the failure rate is constant.
A	$= \dfrac{\text{Up Time}}{\text{Up Time + Down Time}}$	**Availability:** In steady-state operation, this is the probability that the system is up and running over time. For "inherent availability," up time is usually the MTBF and down time is usually the Mean Time To Repair (MTTR) a system. "Noninherent availability" can include complex factors such as standby time, logistic time, and administrative time (also see Chapter 11 on operational availability).

Figure 8.1

Two key areas of reliability

- **Component Reliability (Discretes)**
  - ✓ Resistors, capacitors, diodes, ICs, etc.

- **System Reliability (Hybrids & Assemblies)**
  - ✓ Usually, the whole is equal to the sum of the parts for the failure rate.
  - *Example: Reliability of a light bulb*
    Failure rate = λ system
    λ system = λ filament + λ seal + λ connections
  - ✓ The whole is not equal to the sum of the parts when there is redundancy (double filament inside).

Figure 8.2

Failure rate concepts

- **Time-Dependent Failure Rate** $\lambda(t)$

- **Time-Independent** $\lambda(t) = \lambda$

- **Examples:**
  - ✓ *Time-dependent*
    The failure rate is 1000 FITs at 10.4 years.
  - ✓ *Time-independent*
    The failure rate is constant and is 400 FITs.

- **Instantaneous Failure Rate** (same as hazard rate)

- **Average Failure Rate**

Figure 8.3

Concept of a constant failure rate

$$\lambda(t) = \lambda$$

λ = **F/t Fractional failures/test hours**

= **Number of failures/(total number device x test hours)**

= **Number of failures/total device hours**

λ = **1/Mean Time To Failure = 1/MTTF**

For repairable systems, instead of MTTF, use Mean Time Between Failures.

time period. By quantifying reliability metrics, we can measure and define a target value. Meeting or exceeding the target is then our product reliability objective.

The best way to understand reliability metrics is to first become familiar with the mathematical definition for reliability. Reliability uses a firm set of statistical functions that capture the science and help to measure reliability. The key functions described throughout this chapter are shown in Table 8.1.

In addition to these functional definitions, two key areas of reliability should be defined. These are *system* and *component* reliability. *Component reliability* concerns reliability issues of discretes such as resistors, capacitors, diodes, etc. *System reliability* concerns reliability issues of multiple discretes that make up a unit, such as hybrids, subassemblies, and assemblies. In system reliability, the whole is usually equal to the sum of the parts in terms of the failure rate, unless the system has what is called redundancy (see Figure 8.1).

This chapter does not focus on system reliability. The reader is referred to Chapter 11 on Reliability Predictive Modeling (also see References 1–5) for system reliability modeling. Since systems are built up with components, an understanding of component reliability is needed first to fully appreciate system reliability methods. These two areas of reliability are well defined in industry, to the point where companies have specialized component and/or system reliability engineers.

8.3 Failure Rate Concepts

Before going further with the definitions, it is important to understand some basic concepts about failure rates, since this is the most common reliability metric. The failure rate itself is either time-dependent or time-independent (see Figure 8.2).

As Figure 8.2 indicates, in discussing a time-dependent failure rate, we need to specify the time at which the failure rate is given. The failure rate is also referred to as the instantaneous failure rate (or hazard rate). When the failure rate is given over a time interval, it is referred to as an average hazard rate. Additionally, the failure rate over a specified time interval may be time-independent (see Figure 8.3). When it does not change with time, the hazard rate is constant or simply called a constant failure rate.

When the failure rate is constant, its reciprocal value is the Mean Time To Failure (MTTF). If the hazard rate is not constant, it is generally not identical to the reciprocal of the hazard rate (see References 1–5). Often the term MTTF is used in the context of discrete components or nonrepairable systems. If the system is repairable, the term MTBF (Mean Time Between Failures) is used in place of MTTF (see Chapter 11).

Three common reliability metrics are used for the failure rate: failure per hour, failure per million hours (PPM per hour or year), and the unit of FITs (see Figure 8.4).

▼ **Example 8.1** *Million hour example*

If we have 1% failure (0.01 fractional failure) in 10,000 hours (about 1 year), then

λ = 0.01 fraction fail/10,000 hours
λ = 0.000001 fraction failure per hour

Convert to failures per million (multiply by 1×10^6)
Failure Rate = 1 Failure/million hours

Convert to FITs (multiply by 1×10^9)
Failure Rate = 1000 FITs

Convert to PPM per year (multiply 8760 hours per year) = 8,760 PPM per year

Convert to MTTF = 1/Failure Rate =
1 million hours

Therefore, a one-million-hour MTTF is equivalent to about 1% failure per year [or more precisely, 0.01 fractional failures occur in 10,000 hours (~ 1 year)]. This is 1000 FITs, approximately 10,000 PPM per year. The table below was assembled in a manner similar to this example and may be handy as a quick reference for constant failure rate conversions.

Figure 8.4
Basic reliability metrics

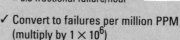

Reliability Metrics

● **Constant failure rate = 1/MTTF**

● **Example: MTTF = 2 hours**
✓ λ = 1/MTTF (failure per hour) = 1/(2 hours) = 0.5 fractional failure/hour

✓ Convert to failures per million PPM (multiply by 1×10^6)
Failure rate = 500,000 Failures/million hours = 500,000 PPM per hour (Note: Multiply by 8760 hours to get ppm per year.)

✓ Convert to FITS (multiply by 1×10^9)
Failure rate = 500,000,000 FITs (Failure in Time)

Table 8.2
Constant failure rate conversion table

FITs	FMH (Fail per 10^6 Hrs.)	MTBF (Hours)	1-Year PPM	1-Year % Failure	2-Year PPM	2-Year % Failure	5-Year PPM	5-Year % Failure	10-Year PPM	10-Year % Failure
1	0.001	1.00E + 09	9	0.0009	18	0.0018	44	0.0044	88	0.009
5	0.005	2.00E + 08	44	0.0044	88	0.0088	219	0.022	438	0.044
25	0.025	40,000,000	219	0.022	438	0.044	1,094	0.11	2,188	0.22
100	0.100	10,000,000	876	0.09	1,750	0.18	4,370	0.44	8,722	0.87
200	0.260	5,000,000	1,750	0.18	3,498	0.35	8,722	0.87	17,367	1.74
400	0.290	2,500,000	3,498	0.35	6,984	0.70	17,367	1.74	34,433	3.44
1,000	1.00	1,000,000	8,722	0.87	17,367	1.74	42,855	4.29	83,873	8.39
2,000	2.00	500,000	17,367	1.74	34,433	3.44	83,873	8.39	160,711	16.07
4,000	4.00	250,000	34,433	3.44	67,681	6.77	160,711	16.07	295,594	29.6
10,000	10.00	100,000	83,873	8.39	160,711	16.07	354,674	35.47	583,555	58.4
40,000	40.00	25,000	295,594	29.56	503,812	50.38	826,573	82.66	969,923	97.0

8.3.1 Mean Time To Failure
Integral Representation

Since the MTTF is the expected time to failure, it is computed with the probability (of failure) density function

$$MTTF = \bar{t} = \int_0^\infty t f(t)dt \qquad (8.1)$$

This integral can be shown to be equivalent to

$$MTTF = \int_0^\infty R(t)dt \qquad (8.2)$$

(when the limit of $tR(t)$ vanishes at large t). In Section 8.4.1, we will describe the exponential reliability model where $R(t) = exp(-\lambda t)$ with a constant failure rate. As an example, the MTTF for this function is

$$MTTF = \int_0^\infty e^{-\lambda t}dt = \frac{1}{\lambda}$$

This is an important result that is discussed in the next section (also see Chapter 11).

8.4 Reliability Models

The failure rate is historically modeled using the traditional bathtub curve shown in Figure 8.5. The curve is modeled after the human mortality rate. *Common reliability failure-rate models fit the bathtub curve.* The regions of the bathtub curve are associated with infant mortality, steady-state operation, and wearout. The infant-mortality period represents a small portion of the shipped population which fails usually in the first year due to possible manufacturing defects that do not immediately show up during screening. The steady-state period represents that portion of the population that fails with a constant failure rate. At end of life, wearout occurs when the failure rate increases in time and the rest of the population fails.

Figure 8.5
Reliability bathtub curve model

Each region can be modeled with a different reliability function. The three main reliability distributions are Weibull, exponential, and log-normal types. The Weibull and log-normal functions are commonly used to model a chang-

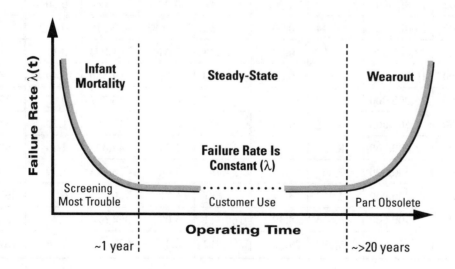

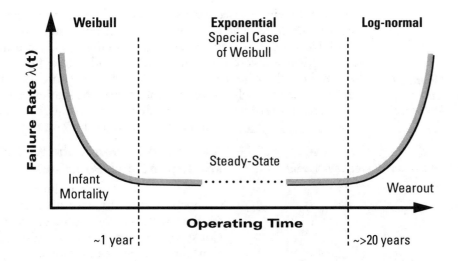

ing failure rate in time while the exponential distribution is used to model a constant failure rate in time (e.g., the steady-state portion of the bathtub curve). The Weibull is most popular for modeling infant mortality while the log-normal function is often used in electronic component reliability to model wearout (see Figure 8.6). There is no definite rule. The deciding factor in choosing a distribution type is to select the distribution function that best fits the data.

One other distribution that is commonly used in reliability statistics is the normal (or Gaussian) distribution, mentioned here for completeness. It is generally not used in modeling the bathtub curve; however, it is used quite often for modeling parameter (parametric or variable) data. This is in contrast to the Weibull, exponential, and log-normal distributions, which are commonly used for catastrophic rather than parametric data. As a summary, the four main reliability models are:

Figure 8.6
Reliability functions come from modeling the bathtub curve

Weibull Distribution
- Can represent any of the three bathtub regions.
- Used mostly in microelectronics for modeling infant mortality.
- Three-parameter model, but only two are commonly needed (third parameter represents a time shift).
- Appropriate for accelerated life tests.

Exponential Distribution
- Constant failure rate.
- Describes only the flat (steady-state) portion of bathtub curve.
- One-parameter model.

Log-Normal Distribution
- Two-parameter distribution.
- Can represent any of the three bathtub regions.
- Used mostly in microelectronics for modeling wearout.
- Replaces time to fail by its logarithm.
- Appropriate for accelerated life tests.

Normal (or Gaussian)
- Two-parameter bell-shaped curve model.
- Used for process monitoring and control charts.

These distributions are the primary focus in this introductory chapter. Further information on these distributions may be found on any of the numerous reliability references currently available (see References 1–8).

In the following subsection, we introduce the functions behind these reliability distributions. The mathematics at this point starts to become less simplified. However, this is the golden age of mathematics, and there are numerous statistical software programs available today to aid in reliability analysis that can help the reader learn to work with these functions. We will provide examples that may help the reader understand the mathematics and show how today's standard software can be invaluable in analysis, compared to years ago when data were more difficult to plot and analyze.

Figure 8.7
The Weibull is a power-law model

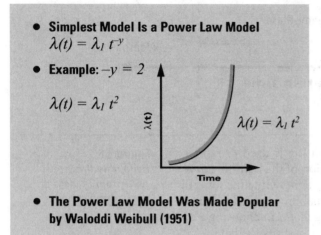

- **Simplest Model Is a Power Law Model**
  $\lambda(t) = \lambda_1 t^{-y}$

- **Example:** $-y = 2$
  $\lambda(t) = \lambda_1 t^2$

  $\lambda(t) = \lambda_1 t^2$

- **The Power Law Model Was Made Popular by Waloddi Weibull (1951)**

8.4.1 Introduction to the Weibull and Exponential Functions

Now that we have introduced these functions in connection with the bathtub curve, it is instructive to demonstrate the connection mathematically. Consider the shape of the bathtub curve in the infant mortality and wearout regions.

If we were asked to model the failure rate wearout region with a function, which would we choose? The simplest is the power-law function, shown in Figure 8.7. The example in the figure illustrates a possible parabolic case for wearout. This power-law function only requires two parameters, with lambda (λ) and y used here. This is the form of an "AT&T" power law representation of the Weibull model (see Reference 1). The original model was made popular by Waloddi Weibull in 1951. *The reader is cautioned that the Weibull model is commonly written in reliability studies with a different parameterization than the AT&T model. The common form of the Weibull model is illustrated in Figure 8.8 and later in 8.10 with $\beta - 1$ as the time power in the failure rate function. This is described in detail in Section 8.4.2. With appropriate substitution, the AT&T and common Weibull models are equivalent. To aid the reader, we have added Appendix A, which provides* conversion between models. In Figure 8.7, we have shifted the wearout axis to time zero. This can be compensated by a shift in the time axis parameter. To do this would require a third parameter, which is often included in what is

Figure 8.8
Modeling the bathtub curve with the Weibull power law

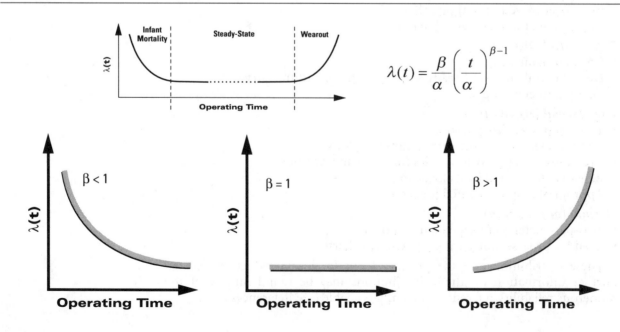

$$\lambda(t) = \frac{\beta}{\alpha}\left(\frac{t}{\alpha}\right)^{\beta-1}$$

termed a three-parameter Weibull model. Actually, any distribution can include a time-shift parameter; using one is only necessary when choosing to fit time-shifted data. Using a third parameter is usually unnecessary as in the case here. As shown in Fig. 8.8, in the common representation of the Weibull model, when $\beta > 1$, we are modeling the wearout region; for $\beta < 1$, we are modeling the infant mortality region. Lastly, when β is identically 1, the failure rate is constant, modeling the steady-state region. Here, the Weibull model is identical to the exponential model.

Figure 8.9
The exponential distribution functions

- **Reliability Function**

$$R(t) = e^{-\lambda t}$$

- **Cumulative Probability Function**

$$F(t) = 1 - R(t) = 1 - e^{-\lambda t}$$

- **Probability Density Function**

$$f(t) = \frac{dF(t)}{dt} = \lambda e^{-\lambda t}$$

- **Failure Rate**

$$\lambda(t) = \lambda$$

▼ **Example 8.2** *The reliability functions of the exponential distribution*

The exponential distribution is mathematically less complex than the other reliability distributions. Therefore, it is instructive to use it as a first example. Here, we find the reliability function $R(t)$ itself. Since the failure rate is constant, from Table 8.1, we have:

$$Fail\ Rate = Constant = \lambda = -\frac{1}{R(t)}\frac{dR(t)}{dt} \quad (8.3)$$

This is a first-order differential equation. The solution is

$$R(t) = e^{-\lambda t} \quad (8.4)$$

This solution may easily be checked by substituting it into the differential equation. The rest of the reliability functions in Table 8.1 are straightforward. Figure 8.9 describes the results. Note that the cumulative probability function can be simplified to λt when λt is much smaller than 1. To show this, one uses a Taylor series expansion where

$$F = 1 - \exp(-\lambda t) = 1 - (1 - \lambda t + Smaller\ order\ terms...) \approx \lambda t \quad (8.5)$$

▼ **Example 8.3** *Using the exponential reliability function*

Problem:
The failure rate of a piece of equipment is constant and is estimated at 10,000 FITs. What is its MTBF? If the Mean Time To Repair this equipment is 1500 hours, what is the inherent availability? If 100 identical units are in the field, estimate the number of units that will fail in the first 6 months and in the interval of 6 to 12 months. If all failed units were returned, what would be the approximate return rate in the first year?

Solution:
The constant hazard rate is 10,000 FITs, this is equivalent to

$$\lambda = 10,000\ FITs\ \frac{10^{-9}/hour}{FIT} = \frac{10000}{10^9 hour} = \frac{10^{-5}}{hour}$$

The MTBF is

$$MTBF = \frac{1}{\lambda} = 100,000\ Hours$$

The inherent availability (also defined in Chapter 11) is

$$A = \frac{MTBF}{MTBF + MTTR} = \frac{100,000}{100,000 + 1500} = 0.985$$

Figure 8.10

Common Weibull model (also see Appendix A)

- **Reliability Function Is Exponential Form**

$$R(t) = Exp[-(\frac{t}{\alpha})^{\beta}]$$

- **Cumulative Distribution Function**

$$F(t) = 1 - Exp\left[-(\frac{t}{\alpha})^{\beta}\right]$$

- **Linear Form of the Cumulative Distribution Function**

$$Ln\left\{Ln\frac{1}{(1-F(t))}\right\} = \beta\ Ln(time) - \beta\ Ln(\alpha)$$

- **Probability Density Function**

$$f(t) = \frac{\beta}{\alpha^{\beta}}t^{\beta-1}Exp\left[-(\frac{t}{\alpha})^{\beta}\right]$$

- **Failure Rate**

$$\lambda(t) = \frac{\beta}{\alpha^{\beta}}(t)^{\beta-1}$$

The probability of failing in the first 6 months (4,380 hours) is Prob. of Failing in first 6 months F(4,380 Hours) = 1 – R(6 months) For the exponential model, R(4,380 Hours) = exp{–4380/100,000} = 0.957, then F(4,380) = 1 – 0.957 = 0.043. Thus, 4.3 units are expected to fail in the first 6 months. This number is rounded up to 5 for a conservative estimate.

The probability of failing in the interval 6 to 12 months is R(6 months) – R(12 months) = 0.957 – 0.916 = 0.041. Thus, 4.1 units are expected to fail in the interval between 6 and 12 months. This number is also rounded up to 5 for a conservative estimate.

The 12-month reliability is 0.916, implying that the return rate is 8.4% in the first year.

8.4.2 Weibull Reliability Functions

Similar to Example 8.2, we can find all the Weibull reliability functions. That is, we can use the power-law model for the failure rate given in Figure 8.8 and solve for the functions in Table 8.1.

This of course leads to some difficult differential equations, whose solutions are provided in Figure 8.10. The reader may wish to show that these solutions satisfy the equations in Table 8.1. Note that the cumulative probability function has also been rearranged in a useful form in Figure 8.10. Later in Example 8.7, a Weibull example is provided in Section 8.6.3 in which this form of the CDF is used.

8.4.3 Normal Distribution Functions

Figure 8.11

Normal distribution model

- **Probability Density Function**

$$f(x) = \frac{1}{\sigma\sqrt{2\pi}}\ exp\left[-\frac{1}{2}\left(\frac{x-p}{\sigma}\right)^{2}\right]$$

- **Mean**

$$\mu = \frac{\sum\limits_{i=1}^{N}x_i}{N}$$
Population Mean

$$\bar{x} = \frac{\sum\limits_{i=1}^{n}x_i}{n}$$
Sample Mean

- **Standard Variance**

$$\sigma^2 = \frac{\sum\limits_{i=1}^{N}(x_i - p)^2}{N}$$
Population Variance

$$S^2 = \frac{\sum\limits_{i=1}^{n}(x_i - \bar{x})^2}{n-1}$$
Sample Variance

Unlike the exponential and Weibull reliability functions, the normal distribution is not commonly used to analyze pass/fail type data. It is used mainly for variable (or parametric) data. Figure 8.11 provides the important functions related to the normal distribution model that are used in this book. As Figure 8.11 indicates, it can be important to differentiate between the population mean and variance, and the sample mean and variance.

▼ **Example 8.4** *Normal distribution analysis of resistors*

Problem:
A sample population of thirty-three resistors are measured with values in ohms of 2.2, 2.3, 2.5, 2.7, 2.9, 2.6, 2.1, 2.4, 2.0, 3.0, 2.55, 2.2, 2.8, 2.4, 2.5, 2.7, 2.9, 2.8, 2.4, 2.6, 2.6, 1.9, 1.8, 3.2, 3.3, 3.3, 2.2, 3.0, 3.2, 2.6, 2.2, 2.8, and 2.9.

Find the sample mean, variance, and standard deviations and plot the frequency distribution.

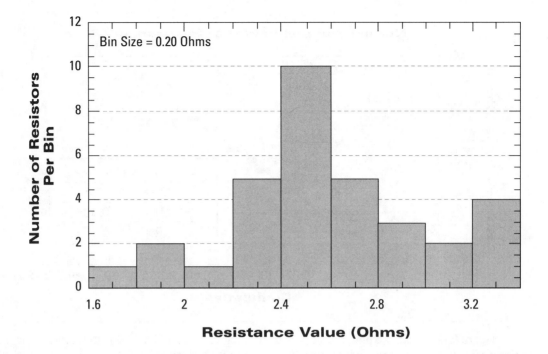

Figure 8.12
Histogram plot of resistance values

Solution:

The sum of the values is 85.55 ohms. Therefore, by using the equation in Figure 8.11 for the sample mean, we have

$$\bar{x} = \frac{85.55}{33} = 2.5924 \ \ ohms$$

To find the variance, we must successively subtract this mean from each value, square the result, and add the total. For example, the first term is

$$(2.20 - 2.531)^2 = 0.1096$$

The sum of these values is 5.0406 ohms. This is then divided by $n - 1 = 32$ values yielding

$$S^2 = 0.15752$$

The standard deviation is then found by taking the square root of the variance resulting in S = 0.3969. Note, these values can be obtained using the Excel functions in Table B.1.

Frequency distributions can be displayed in a number of ways. We recommend the use of a good statistical software program when plotting normal distributions. We provide two examples that are commonly plotted from such software. The first is the popular histogram plot (see Figure 8.12).

However, it is often difficult to tell using histogram plots, whether or not the data has the bell-shaped normal distribution, as this plot is highly dependent on histogram bin size. An alternate plot that is commonly used in place of the normal histogram plot is a normal probability plot. In this case, the resistance values are plotted versus the cumulative probability values. Note that sometimes the x-axis can also be plotted in terms of standard normal deviations rather than the cumulative probability values (this varies when using software for plotting). This is displayed in Figure 8.13. It is easy to see the advantages. If the data are normally distributed, values will fall on a straight line which can easily be assessed. The data can also be fitted using linear regression as shown in Figure 8.13. In

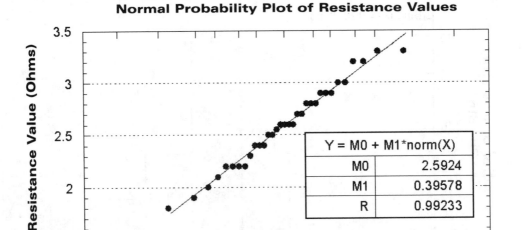

Normal Probability Plot of Resistance Values

Y = M0 + M1*norm(X)	
M0	2.5924
M1	0.39578
R	0.99233

Figure 8.13
Normal probability plot of resistance values

such a plot, the slope found in regression analysis yields the standard deviation and mean intercept. From Figure 8.13, the results indicate the mean and standard deviation values are in agreement with our results, with the high regression coefficient R^2 of 0.992 (see the table in Figure 8.13) indicating the degree of normality.

8.4.4 The Log-Normal Reliability Function

The log-normal Probability Density Function (PDF) is analogous to the normal PDF with the exception that we take the logarithm of the parameter values. This is shown in Figure 8.14, where the important log-normal functions are provided. An example of using these functions is given later in this chapter.

8.5 Reliability Objectives and Confidence Testing

Figure 8.14
Important log-normal reliability functions

Now that we have introduced reliability metrics and functions, understanding their usage in product reliability testing is important for practical applications. In today's competitive marketplace, reliability testing is often customer-driven. Some customers know what they want, while others simply want a bottom-line reliability that their product will work over life hazard conditions throughout its useful life. The goals are the same, but the manner in which one "proves-in" a product may vary. From a statistical point of view, a quantitative assessment will start with the product's reliability objective, then statistically demonstrate this objective at a certain confidence level.

Reliability objectives will vary depending upon a product's capability. Figure 8.15 provides common product reliability failure rate objectives (in FITs) for plastic ICs, hybrids, and assemblies. Reliability testing is somewhat of an inexact science. The best that can be done is to use a statistically meaningful sample size to make inferences about the population with a certain level of confidence (Figure 8.16). The samples that are tested must represent the population, or confidence becomes uncertain. What then is confidence?

- **Probability Density Function**

$$f(t) = \frac{1}{\sigma\, t\sqrt{2\pi}} \quad \exp\left\{-\frac{1}{2}\left(\frac{\ln t - \ln t_{50}}{\sigma}\right)^2\right\}$$

- **Cumulative Probability Function**

$$F(t) = \frac{1}{\sigma\sqrt{2\pi}}\int_0^t \frac{dx}{x}\exp-\frac{1}{2}(\frac{\ln(x/x_{50})}{\sigma})^2 = \frac{1}{2}\left[1 + erf\left(\frac{\ln(t/t_{50})}{\sqrt{2}\sigma}\right)\right]$$

- **Failure Rate**

$$\lambda(t) = \frac{f(t)}{1 - F(t)}$$

- **Median = t_{50}**

- **Shape Parameter**

$$\sigma = \ln(t_{50}/t_{16})$$

Confidence Test
A statistically significant test to demonstrate a specific reliability objective at a certain confidence level

Reliability Objective	Plastic ICs (FITs)	Hybrid (FITs)	Assem. (FITs)
1	5	400	400
2	50	1000	1000
3	100	2000	40000
4	400	4000	10000

There are two basic types of confidence: engineering confidence and statistical confidence. To some extent, we often use a certain amount of engineering judgment in choosing a representative sample, frequently assuming that measurements are accurate, trusting that the test(s) are run properly, and so forth. These are common factors that are often judged from an engineering point, due to time and cost restrictions. Statistical confidence, however, is integral to sample planning and demonstrating that a reliability objective can be met with a certain level of confidence. The confidence level (or confidence coefficient) is usually expressed by a percentage. This is commonly associated with a confidence interval (or limits) as shown in Figure 8.17.

For example, if we are trying to estimate the population mean from a sample set, we make a number of measurements to estimate its value. The obtained value is a point estimate. We have a certain amount of uncertainty in the actual population mean. Our uncertainty can be quantified statistically by a confidence interval. This depends on the sample size and what level of confidence we use. Statistically, the confidence interval indicates the degree of uncertainty in our measured estimated value for the population

Figure 8.15
Demonstrating a reliability objective at a certain level of confidence

Figure 8.16
What is confidence?

- **Engineering confidence is largely a matter of judgment and experience.**
  - ✓ I am confident that this design is as reliable as our previous design.

- **Statistical confidence is used to make inferences about a population, given data from a sample.**

Figure 8.17
Statistical confidence interval

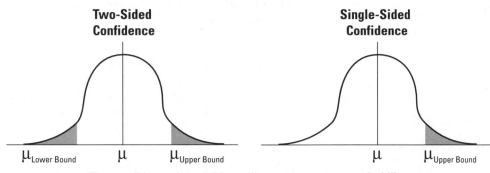

The confidence interval gives the range of values between which an observation is expected to lie within a given probability.

Two-Sided Confidence

$\mu_{Lower Bound}$ μ $\mu_{Upper Bound}$

Single-Sided Confidence

μ $\mu_{Upper Bound}$

The confidence interval here illustrates percent probability that an observation will fail within the unshaded range, and this degree of confidence is referred to as the percent confidence level.

Figure 8.18
Reducing the confidence interval

mean. As shown in Figure 8.18, as we increase the number of observations (larger sample size), our statistical interval becomes smaller at the particular confidence level we have chosen.

- A larger representative sample size narrows the interval for the estimate of population's mean for the same level of confidence.

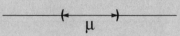

- Longer test time gives a better estimate of true fractional failure. The interval gets smaller as time increases for the same confidence level.

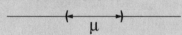

8.6 Parametric and Catastrophic Methods

Because we are mainly interested in either catastrophic or parametric problems, appropriate statistics should follow with parametric or catastrophic statistical methods. This is described in Figure 8.19. Of course, a number of in-depth topics can be covered in each. For the purpose of this book, we cover a few important aspects in this section.

8.6.1 Parametric Overview

Statistical confidence, as it relates to normal parametric applications, is essentially depicted in Figure 8.17. In actual practice, it is useful to know a confidence interval about a measured mean as shown in Figure 8.18. To do this, we take a sample of size n from the population of size N. The sample's measured mean is $\bar{x}$ and standard deviation is S.

We wish to estimate the population mean μ having a standard deviation of σ, with a certain level of confidence. The best we can do is to estimate the confidence interval given by

$$\bar{x} - z_{\alpha/2} \frac{\sigma}{\sqrt{n}} < \mu < \bar{x} + z_{\alpha/2} \frac{\sigma}{\sqrt{n}} \qquad (8.6)$$

where $z_{\alpha/2}$ is the Z value of a standard normal distribution leaving an area of $\alpha/2$ to the right of the Z value (see Reference 4). The Z value for any random variable x is $Z = (x - \mu)/\sigma$. For small samples (n < 30) from an approximate normal population with unknown variance, the confidence interval for μ is given by

$$\bar{x} - t_{\alpha/2} \frac{s}{\sqrt{n}} < \mu < \bar{x} + t_{\alpha/2} \frac{s}{\sqrt{n}} \qquad (8.7)$$

Figure 8.19
The two main kinds of statistics to analyze data and estimate confidence.

where $t_{\alpha/2}$ is the t value with $v = n - 1$ degrees of freedom, leaving an area of $\alpha/2$ to the right (see Reference 4). The confidence equation is illustrated in the following example.

- **Parametric Statistics**
  - ✓ CPK analysis (design maturity testing, etc.)
  - ✓ Parameter analysis (process reliability)
  - ✓ Accelerated testing parameter data
  - ✓ Parametric confidence

- **Binomial Statistics (Pass-Fail, Go/No-Go)**
  - ✓ Catastrophic pass-fail analysis
  - ✓ Accelerated test sample size planning
  - ✓ Binomial confidence

▼ **Example 8.5** *Parametric confidence interval about the mean*

Problem:
In Example 8.4, we estimated that the mean value of a sample of 33 resistor measurements was 2.592 ohms with a standard deviation of 0.397 ohms. We would like to find the 95% confidence interval about this mean for the actual parent population.

Solution:
Since the sample size is large (>30), the standard deviation σ can be approximated by $S = 0.397$. We are interested in the 95% confidence interval. This means that $\alpha = 1 - 0.95 = 0.05$, and $\alpha/2 = .025$. The

Z value estimate rather than the *t* value can be used since we are close to a sample size of 30. That is the *Z* value that leaves an area of 0.025 (= 0.05/2) to the left and therefore an area of 0.975 to the right is $Z_{0.025} = 1.96$. Note that for this sample size, the *t* value happens to be the same, *t* statistics will provide the same results. This *Z* or *t* value can be found from a statistical table for $\alpha/2$. Alternately, one can use a software package like Microsoft® Excel. An Excel example for this problem is provided in Appendix B. With this value, the 95% confidence interval is

$$2.592 - (1.96)\ (\frac{0.397}{\sqrt{33}}) < \mu < 2.592 + (1.96)\ (\frac{0.397}{\sqrt{33}})$$

which reduces to

$$2.727 < \mu < 2.457$$

Therefore, there is a 95% probability that an observation for a resistance value will fall in this range, and the degree of confidence is referred to as the 95 percent confidence level.

8.6.2 Central Limit Theorem and Cpk Analysis

One of the key theorems in statistics for normal distribution is the Central Limit Theorem, which is described in Figure 8.20. The theorem is important for these studies as it applies to sampling. It states that sampling means follow approximately the normal distribution even if the underlying distribution is not normal. Most of the time, we are not fortunate enough to know the variance of the population from which we select our random samples. According to Reference 9, "For samples of size *n* > 30, a good estimate of σ^2 is provided by calculating s^2. If the sample size is small (*n* < 30), the values of s^2 can fluctuate considerably from sample to sample." Thus, it is a good rule of thumb when performing statistical measurements to use samples above 30.

▼ **Example 8.6** *Cpk analysis*

One important area in sampling is process capability analysis. In obtaining the Cpk (capability) Index, a normal distribution is assumed for the test measurements. The purpose of process capability analysis is to verify that all key parameter measurements remain within the process capability indices limits. This is the Cpk index. In Cpk analysis, a statistically meaningful sample is chosen. The Central Limit Theorem illustrates the validity of sampling from a normal population. In Cpk analysis, the variance plays a key role. Since in practice we usually do not know the population variance and estimate it from the sample, we seek to minimize its fluctuation from sample to sample. Therefore, we recommend using a nominal sample size of at least 30 units as described in Figure 8.20 and Reference 9. The statistical distribution should optimally be a preferred Cpk value greater than 1.5. Figure 8.21 provides the statistical definitions for the

Figure 8.20
Parametric statistics: The Central Limit Theorem

Central Limit Theorem

- Sampling means follow approximately the normal distribution even if the underlying distribution is not normal. If we have a sample size of *n* from a population with a mean μ and a finite variance σ^2, with an increase in sample size *n*, the distribution of sample means approaches a normal distribution with a mean μ and variance σ^2/n (see Reference 4).

- Rule of Thumb: The normal approximation in the Central Limit Theorem will be good if *n* > 30, regardless of the shape of population. If *n* < 30, the approximation is good only if the population is not too different from a normal population. For a normal population, the distribution will follow a normal distribution exactly, no matter how small the size of the samples (see References 4 and 9).

Figure 8.21
Cpk analysis

Cpk is a measure of meeting specified limits and target

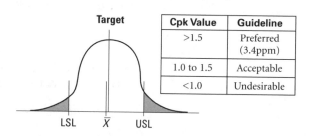

Cpk Value	Guideline
>1.5	Preferred (3.4ppm)
1.0 to 1.5	Acceptable
<1.0	Undesirable

$$\sigma\ Capability = Min\left[\frac{\overline{X} - LSL}{\sigma}\ ,\ \frac{USL - \overline{X}}{\sigma}\right]$$

$$Cpk = \frac{\sigma\ Capability}{3}$$

Cpk Value	σ Capability	2-Sided PPM	2-Sided Normal Percent	1-Sided PPM	1-Sided Normal Percent
2.00	6	0.002	99.9999998	0.001	99.99999990
1.667	5	0.6	99.99994	0.3	99.99994
1.50	4.5	6.8	99.99932	3.4	99.99966
1.333	4	63	99.994	32	99.997
1.166	3.5	465	99.95	233	99.98
1.00	3	2,700	99.73	1,350	99.87
0.833	2.5	12,419	98.76	6,210	98.76
0.667	2	45,500	95.45	22,750	97.73
0.500	1.5	133,615	86.64	66,807	93.32
0.333	1	317,311	68.27	158,655	84.13

Table 8.3

Relationship between the sigma (σ) capability, Cpk index, and yield

Cpk indices. Table 8.3 provides an overview of the statistical relationship between the Cpk index, process σ capability, and yield. In Table 8.3, both one-sided and two-sided normal distribution values are provided. These values can be found from tables. Alternately, one can use software like Microsoft® Excel. An Excel table example is provided in Appendix B. Often, we need to consider whether or not we have a one-sided or two-sided specification. Note that the Cpk value is insensitive to one-sided or two-sided specifications. Thus, the index may not accurately portray yield information. As a specific example, consider the results in Example 8.4. In this example, 33 resistors were measured, a total which is slightly above our minimum recommended sample size. The mean resistance value was measured at 2.592 ohms. If the upper specified limit (USL) and the lower specified limit (LSL) are defined as USL = 4.2 ohms and LSL = 1.1 ohms, then the mean distance to the specification limits are

$$USL - Mean = 1.61 \ ohms \quad Mean - LSL = 1.49 \ ohms$$

Therefore, the Mean-LSL is the minimum value as shown in Fig 8.21. From Example 8.4, sigma(σ) is 0.397, then the sigma capability of this process is

$$\sigma \ Capability(LSL) = Z_{LSL} = Mean - LSL/\sigma = 1.49 \ ohms/0.397 = 3.76$$

and the Cpk index is

$$Cpk = 3.76/3 = 1.25$$

Generally, we only look at the sigma capability for the worst tolerance side. It is instructive to find it for the upper limit as well. This is

$$\sigma \ Capability(USL) = Z_{USL} = Mean - USL/\sigma = 1.61 \ ohms/0.397 = 4.06$$

With this, we can find the anticipated resistor proportion out of specification. Using a one-sided normal analysis, the PPM value in a standard table (see Appendix B) for 3.76 sigma capability is 85 PPM, while for 4.06 sigma capability process it is 24.5 PPM. Therefore, a total of 109.5 PPM resistors are anticipated to fall out of this process. The total anticipated resistor yield is 99.98905% (= 1 − 109.5 PPM).

8.6.3 Catastrophic Analysis

There are two types of reliability failures: parametric and catastrophic. As we just discussed, a parametric failure occurs when a device's parameter values exceed the customer's parametric limits. In the above Cpk example, these were the USLs and LSLs. In catastrophic analysis, we are only concerned with a pass/fail criterion. We commonly use the Weibull, log-normal, and exponential distributions to analyze pass/fail catastrophic data and the normal distribution for parametric data analysis. However, each distribution can apply in either the catastrophic or parametric case depending on the data; only the analysis changes. In the catastrophic case, we are interested in the failure history rather than the parameter values or parameter shifts. Catastrophic failure information usually includes the time of failure for that fractional portion of the population that has failed. For example, life test data on a sample population can be used to estimate the cumulative distribution function for the population. If the exact failure times for all the units tested are known or estimated, a probability plot can be obtained. The data procedure is as follows:

- Failure times for n failures are ranked, arranged from smallest to largest.
- Cumulative probability plotting position indicating $F(t)$ are assigned for each i_{th} failure for F_i either by expected $i/(n + 1)$, midpoint $(i - 0.5)/n$, or median $(i - 0.3)/(n + 0.4)$ plotting position multiplied by 100 to obtain percent.
- The cumulative probability plotting values are usually transformed so that the data can be plotted in a form that can be fitted.

Censored Data

Reliability studies often result in less-than-complete or censored data. For example, devices can be removed during a test, a test may not be run to completion, or exact failure times are not known. When life test data are analyzed, some units are unfailed, and their failure times are known only to be beyond their present running times. Such data are said to be *censored on the right (failure time > t_o)*. A failure time known only to be before a certain time is said to be *censored on the left (failure time < t_o)*. However, if a failure time is known to be within an interval when it is not continuously monitored, it is said to be *interval censored (t_o < failure time < t_1)*. If all units are started together on test and the data are analyzed before all units fail, the data are singly censored. Data are *multiply censored* if units have different running times intermixed with the failure times. Time-censored data are also called Type 1 censored and are the most common type of censored data. There are a number of methods used in data analysis for singly and multiply censored data (see References 3 and 7). One example is provided below.

▼ **Example 8.7** *Weibull and log-normal analysis of low-noise amplifier life test failures*

Problem:

Life test data for low-noise amplifiers (LNAs) are listed in Table 8.4. These data are from a second experiment similar to the one previously plotted using the log-normal distribution in Chapter 6 (see Figure 6.7). We would like to analyze this LNA life test data using a Weibull analysis. Devices were put on life test at 200°C and 250°C. Failure times were interval censored, every 100 hours at 200°C and every 10 hours at 250°C. Using the Weibull model, find α and β for both the 200°C and 250°C data. Give the failure rate at any time. Then using this information, estimate the MTTF at each temperature. Replot the data in the log-normal case and compare it to the results in Chapter 6 and to your Weibull analysis.

Rank K	Cum Failures i	Plotting Position Fs = i/(n + 1)	Ln {–Ln(1 – Fs)}	Failure Times at 200°C (Hours)	LN (Failure Time) at 200°C	Failure Times at 250°C (Hours)	LN (Failure Time) at 250°C
1	1	0.063	–2.732	8800	9.083	260	5.561
2	2	0.125	–2.013	9100	9.116	290	5.669
3	3	0.188	–1.569	9200	9.127	360	5.886
4	4	0.250	–1.246	9500	9.159	390	5.966
5	5	0.313	–0.980	9600	9.170	450	6.109
6	6	0.375	–0.755	9900	9.200	510	6.234
7	7	0.438	–0.551	10000	9.210	610	6.414
8	8	0.500	–0.367	10500	9.259	720	6.579
9	9	0.563	–0.189	11000	9.306	740	6.607
10	10	0.625	–0.019	11100	9.315	750	6.620
11	11	0.688	0.153	12000	9.393	790	6.672
12	12	0.750	0.327	13000	9.473	800	6.685
13	13	0.813	0.517	13500	9.510	810	6.697
14	14	0.875	0.732	14000	9.547	830	6.721
15	15	0.938	1.023	15000	9.616	900	6.802

Table 8.4

Life test data arranged for plotting rank

Solution:

Each failure is ranked with the Mean Time To Failure data arranged in ascending order as shown in Table 8.4. In this analysis, the expected plotting position is used {i/(n + 1) × 100}. The ranking is performed to estimate the sample's cumulative probability function denoted as F_s. Note that the rank happens to correspond to the number of failures in this case. This is because only one failure was observed at each failure time, which is not always the case (for example, see Example 9.11 in Chapter 9). Since the failure times were interval censored, the upper interval point was used. Additionally, all 15 units failed. However, in most analyses not all failure times are obtained. A linearized form of the Weibull cumulative probability function is provided in Figure 8.10. Since we wish to fit data with a straight line best fit analysis, we linearize the data by plotting $Ln\{-Ln(1 - F_s)\}$ values versus Ln(time). Then the data are plotted and fitted using a linear fit as shown in Figure 8.22. The results of the fits are displayed in the figure. Note that the regression coefficients are 0.93 and 0.97 for the 200°C and 250°C data, respectively. This indicates that the Weibull analysis is reasonable. We can compare the regression coefficient for the Weibull analysis to that of the log-normal data in Figure 6.7 which was 0.97 and 0.94 for the 200°C and 250°C data, respectively. There is no clear indication that either the log-normal or Weibull distribution fits the data better. Next, we proceed to make a comparison between the Weibull function given in Figure 8.10 where

$$Ln\{-Ln(1 - F(t))\} = \beta \ Ln(time) - \beta \ Ln(\alpha) \qquad (8.8)$$

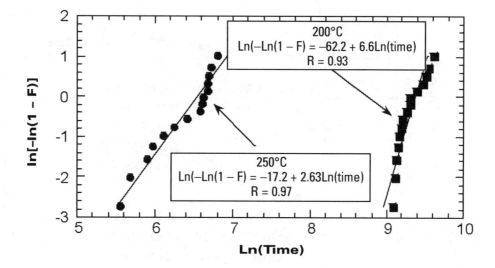

Figure 8.22
Weibull analysis for example

and the linear fit $Y = B + MX$ of form obtained in Figure 8.22

$$Y = -62.2 + 6.63\ Ln(time) \quad \text{and} \quad Y = -17.2 + 2.63\ Ln(time)$$

for the 200°C and 250°C data, respectively. Note: for most accurate results, the regression should be performed on the uncertainty parameter of time, not plotting position. By comparison, where $Y = Ln\{-Ln(1 - F)\}$ and $X = Ln(time)$, we have

$$\beta = 6.6\ for\ the\ 200°C\ fit \quad \text{and} \quad \beta = 2.63\ for\ the\ 250°C\ curve$$

Also we have $-62.2 = -\beta\ \ln(\alpha)$ for the 200°C data and similarly $-17.2 = -\beta\ \ln(\alpha)$ for the 250°C curve. These values give $\alpha = 11,801$ hrs for the 200°C curve and $\alpha = 694$ hrs for the 250°C curve. The failure rates are obtained using the expression in Figure 8.11 as

$$\lambda(t) = \frac{\beta}{\alpha}(t)^{\beta-1} \qquad (8.9)$$

giving

$$\lambda(t) = 6.7x10^{-27}\ t^{5.63} \quad \text{and} \quad \lambda(t) = 8.8x10^{-8}\ t^{1.63}$$

for 200°C and 250°C, respectively. The reader may wish to find the common Weibull model parameters using conversion Table 8.A.2 in Appendix A.

To find the MTTF, we solve the cumulative distribution function above (or see Table 8.A.1) for t giving

$$t = \alpha\left(Ln\left\{\frac{1}{1-F}\right\}\right)^{\frac{1}{\beta}} \qquad (8.10)$$

To obtain the MTTF, we insert $F = 0.5$ and the values for λ_1 and α to obtain 11,376 and 610 hours for 200°C and 250°C data, respectively.

To compare these results to that of a log-normal analysis, we replot the data on a normal probability plot shown in Figure 8.23. Here, we have chosen to plot the *ln*(failure time) values instead of the failure time values. Either can be plotted. In this case, the failure time axis can be linear compared with the failure time axis in the first experiment in Figure 6.7. The linear least-

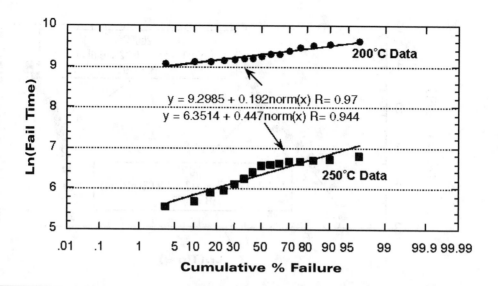

Figure 8.23
Normal probability plot for the example

squares best fit shows that the MTTF times are 10,921 (= exp(9.299) and 573 (= exp(6.35)) for the 200°C and 250°C data, respectively. Note that the log-normal slopes are 0.192 and 0.447. These values compare favorably to the first experiment in Figure 6.7.

In comparison to the MTTF Weibull plot, we note that the log-normal regression coefficients are close to being reversed for the 200°C and 250°C data. Thus, there is no justification that one analysis is any better that the other.

The reader might note that both β and α change with temperature. How would we predict these values at other temperatures? In order to fully analyze the data, one must introduce the temperature Arrhenius function (see Chapter 9) into the Weibull model. This becomes a three-dimensional problem since the Arrhenius function introduces another parameter that must be simultaneously fitted along with the two Weibull parameters. This type of a problem requires multivariable analysis, which is outside the scope of this book.

▼ **Example 8.8** *MTTF log-normal confidence interval*

Problem:
Obtain the 90% confidence limits around the MTTF value for the 200°C data in the log-normal plot in Figure 8.23.

Solution:
In Figure 8.23, the MTTF is exp(9.30) = 10,921 hours with a standard deviation of 0.167. The confidence interval around a mean is given in Section 8.6.1. When the sample size is less than 30, the interval recommended is in terms of the t statistic. Substituting the transformed log-normal MTTF value into the equation yields

$$9.3 - t_{\alpha/2}\frac{0.167}{\sqrt{15}} < \mu < 9.3 + t_{\alpha/2}\frac{0.167}{\sqrt{15}}$$

The *t* statistic for 90% confidence (found in most statistics books) can be obtained with the aid of an Excel equation. This is provided in Appendix B, Table 8.B.1, and the value given in this table is 1.7613. Inserting this above, the 90% confidence interval ranges from exp(9.223) to exp(9.374) or 10,127 to 11,778 hours.

▼ **Example 8.9** *Mixed failure analysis*

When different failure mechanisms occur during life testing for the same sample set, results should be treated separately. The failure mechanisms may be accelerated at different rates and require independent treatment. One such method is the product-limited method described by Kaplan and Meier (see Reference 11).

This can be viewed using the conditional probability where the probability for survival, $P_{s,n}$, for a particular failure mechanism to time t_n (failure time) is equal to the cumulative probability of survival $P_{s,n-1}$ to the previous time t_{n-1}, the measurement time for the previous failure, multiplied by the probability of survival, P_s, from time t_{n-1} to t_n or

$$\text{Cum } P_{sn} = \text{Cum } P_{s,n-1} \, P_s \qquad (8.11)$$

As an example, Table 8.5 shows life test data of mixed electrical and mechanical failure modes. The initial failure has a mechanical failure mode at 3 hours. The next mechanical failure, occurring at 10 hours, does not occur until after two other electrical failures. Therefore, the probability at the 10-hour point for the mechanical failure mechanism can only be measured out

Table 8.5

Life test data with mixed failure mechanisms

Number on Test	Time (Hours)	Mech Failures	Elec Failure	Mech Survive	Mech Cum Fail	Elec Survive	Elec Cum Fail
20	3	1		19/20 = 0.96	4		
19	5		1			18/19 = 0.94737	5.2632
18	8		1			17/18 = 0.94444	5.5556
17	10	1		(.96)(16/17) = 0.9	10		
16	20		1			(.94)(15/16) = 0.89	11
15	30		1			(.89)(14/15) = 0.83	17
14	40	1		(.9)(13/14) = 0.84	16		
13	100	1		(.84)(12/13) = 0.78	22		
12	320		1			(.83)(9/12) = 0.62	38
11	320		1				
10	320		1				
9	500	1		(.78)(7/9) = 0.61	39		
8	500	1					
7	900		1			(.62)(7/8) = 0.54	46
6	1200	1		(.61)(5/6) = 0.51	49		
5	1300		1			(.54)(4/5) = 0.43	57
4	2000	1		(.51)(3/4) = 0.38	62		
3	2100		1			(.43)(2/3) = 0.29	71
2							
1							

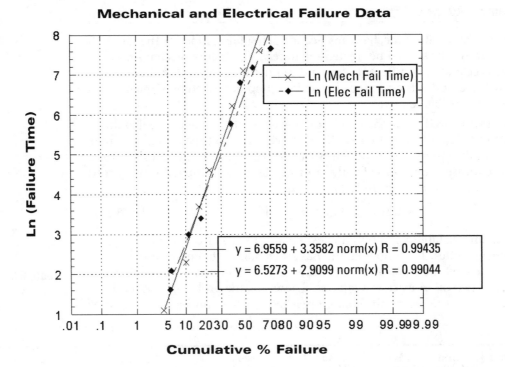

Mechanical and Electrical Failure Data

Figure 8.24
Two different failure mechanisms observed on life test

of 17 units, 16/17 times the previous survival probability of 0.96, yielding a probability of survival at that time of 0.9. The plotting position then becomes 10% failure at the 10-hour measurement point. Figure 8.24 shows the corresponding log-normal probability plot.

▼ **Example 8.10** *S-shaped data analysis*

In Example 8.9, the distribution is made up of two failure mechanisms. It would be difficult to see the two modes if the data had been plotted without separating out failure mechanisms. This can occur when both modes fail in the same time frame throughout the test. Often in life-testing, a subpopulation and main population are observed occurring at two distinct time frames. The subpopulation (sometimes called "freaks") shows up at early test times compared to the main failure population. The characteristic is that of an S-shape as illustrated by the bimodal life test data (marked with X) in Figure 8.25. This behavior can occur from the same failure mode. For example, cracks in semiconductor components followed by metal diffusion in the cracks and eventual shorting of the junction often appear in an S shape. Perhaps, because the failure mode is compounded by two mechanisms, the failure rate is compounded in time and takes on this classical S characteristic. Recognizing this, one sees a distinct separation at an inflection point in the S-shaped data. In Figure 8.26, the inflection point is taken at 30%. This divides the total population into subpopulation (30%) and main population (70%) groups. The life test data for Figure 8.25 are displayed in Table 8.6 obtained from 15 devices. The observed failure times are listed in Table 8.6.

The subpopulation is then assessed below this inflection point. The values for the subpopulation in column 3 are renormalized by the maximum 30% value. For example, the point at 51.2 hours and 4.55% is transposed as 4.55%/0.3 × 100 = 15.15%. The point is now plotted as part of the

Rank (i − 0.3)/(n + 0.4)	Population Fail Time	Sub Pop Fail Time	Main Pop Fail Time	Population Ln (Time)	Sub Pop Ln (Time)	Main Pop Ln (Time)
0.74211			159.3			5.07
4.5455	51.2			3.94		
10.019			273			5.61
11.039	55.4			4.01		
15.152		51.2			3.94	
17.532	78.4			4.36		
19.295			313			5.75
24.026	86.8			4.46		
28.571			318			5.76
30.52	159			5.07		
36.797	273	55.4		5.61	4.01	
43.51	313			5.75		
50	318			5.76		
58.442		78.4			4.36	
80.087		86.8			4.46	

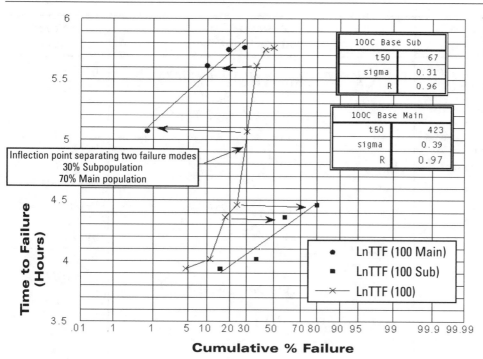

Table 8.6
Life test data and the renormalized groups

Figure 8.25
Life test data displaying an S shape replotted as sub- and main groups

Figure 8.26
Estimating samples to demonstrate a reliability objective of 100 FITs

---

Pass/Fail Testing — How Many Devices for How Long?

- **How do I *demonstrate* a reliability objective of 100 FITs?**

  (Recall 1 FIT = 1 failure/1×10^9 hours, i.e., 100 FITs = 1 failure/1×10^7 hours)

- **Possible Tests (1 failure allowed)**
  - ✓ Test 1 device for 1×10^7 hours (~ 1,100 years)
  - ✓ Test 100 devices for 1×10^5 hours (~ 11 years)
  - ✓ Test 100 devices for 1,000 hours (~ 1 month) and **accelerate time** by a factor of **100**

---

subpopulation at 51.2 hours and 15.15%. The other points in the subpopulation are found similarly. Column 4 is the transposed values for the main population. These are the points above the inflection point. This population is renormalized by subtracting the inflection point value and then dividing by 70%. For example, the 50%, 318-hour point is transposed as (70% − 50%)/0.70 = 28.6%. This is then plotted as part of the main population at 318 hours and 28.57%. The other points for the main population are found similarly.

8.6.4 Reliability Experiments with Catastrophic Data

The pass/fail type of reliability experiments can either be a simple demonstration, an investigative test, or a statistically significant test. The objective is to estimate the success (or failure) probability for the total population to pass (or fail) the test, based on the results from a sample set. The reliability aspect usually infers that the test is performed over time. Figure 8.26 illustrates a demonstration test to show that devices can meet a 100-FIT objective. As pointed out in the figure, the test design depends on sample size, test time, and as the last scenario indicates, the environmental acceleration factors, which will be discussed later. In the example, the failure rate is treated as constant.

Table 8.7
Confidence intervals for pass/fail data

Method	Usage	2-Sided 100 q% Limit		1-Sided 100 q% Limit
Normal Approx.	Ideal: $n - > \infty$ $p - > 0.5$ Guide: $n - y > 5$	$\underline{P} \cong \tilde{P} - Z_{(1+q)/2}\{\tilde{P}(1-\tilde{P})/n\}^{0.5}$ $\overline{P} \cong \tilde{P} + Z_{(1+q)/2}\{\tilde{P}(1-\tilde{P})/n\}^{0.5}$		Replace $(1+q)/2$ by q
Poisson (Chi-square) Approx.	Ideal: $n - > \infty$ $p - > 0$ (small) Guide: $10y < n$	$\underline{P} \cong \dfrac{\chi^2\{(1-q)/2;2y\}}{2n}$ $\overline{P} \cong \dfrac{\chi^2\{(1+q)/2;2y+2\}}{2n}$		Replace $(1-q)/2$ by $(1-q)$ Replace $(1+q)/2$ by q
Standard Limits	Any	$\underline{P} \cong \{1 + (n - y + 1)/y$ $F[(1+q)/2;2n-2y+2,2y]\}^{-1}$ $\overline{P} = (1 + (n-y)\{(1+y)$ $F((1+q)/2;2y+2,2n-2y)^{-1})^{-1}$		Replace $(1+q)/2$ by q

Table Key: y = number of failures, q = fractional confidence of interest, n = sample size,
$\tilde{P}$ = y/n = point estimate of fraction failed, $\overline{P}$ = upper bound, $\underline{P}$ = lower bound,
$x^2(q,v)$ = 100qth% = chi-square distribution with v degrees of freedom
F = F – distribution

In a statistically significant test, sample sizes, test times, and when applicable, acceleration factors are tied to statistical confidences. The easiest way to design a statistically significant pass/fail test is by using confidence intervals for the experiment. In general, this test implies that we perform an experiment on n samples that are representative of the total population. The binomial outcome is simply that samples will either pass or fail. Furthermore, samples should be statistically independent of each other such that the outcome of one sample will not affect the outcome of another. Under these assumptions, Table 8.7 provides an overview of three types of confidence intervals that may be used for designing pass/fail tests (see Reference 7).

▼ **Example 8.11** *Confidence interval example*

Problem:

To add a measure of statistical significance to the reliability demonstration test example in Figure 8.26, find the single-sided upper 90% confidence bound on the failure rate for the test described in that figure.

Solution:

In this test, the number of failures y is 1, the fractional confidence q is 0.9, and the sample size n is 100. Referring to Table 8.7, we see that (usage column) $n/y > 10$, so we can use the chi-square approximation. From Table 8.7, we make the appropriate substitution for the single-sided upper confidence bound estimator on the fractional fail where

$$\overline{P} \cong \frac{\chi^2\{q;2y+2\}}{2n} = \frac{\chi^2\{90\%;2*1+2\}}{2*100} = \frac{\chi^2\{90\%;4\}}{200} = \frac{7.78}{2*100} = 0.039$$

The value $\chi^2(0.9,4) = 7.78$ may be found in statistical tables or obtained in Excel as shown in Appendix B, Table 8.B.1. In Figure 8.26, for $n = 100$, the test time is 100,000 hours (note that the accelerated test time is actually only 1,000 hours), then assuming a constant failure rate, the upper bound is

$$\overline{\lambda} = \frac{0.039}{100,000} = 0.00000039 \ failures/hr = 390 \ FITs$$

The test indicates that we have demonstrated 100 FITs. However, since we have tested only a sample, the statistical outcome indicates the degree of uncertainty about the 100-FIT point estimate. This is illustrated in Figure 8.27. The single-sided upper confidence bound is 390 FITs, and our degree of confidence is 90%. Therefore, there is a 90% probability chance that any observation will be no worse than this upper bound.

Finally, it should be mentioned that the confidence interval can be used as a hypothesis test. When we design a binomial test, we are actually estimating a sample size based on a pass/fail hypothesis criterion. For instance, in the above example, a hypothesis test design would be based on the single-sided upper 390-FIT criterion at the 90% confidence level using the chi-square estimator. We would accept or reject the hypothesis based on the outcome. Failing to reject the hypothesis indicates that the outcome is within the confidence interval, resulting from no more than one failure out of the 100 samples tested.

Figure 8.27
Reliability test results for one failure out of 100 samples

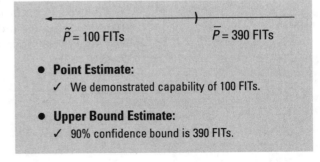

$\widetilde{P} = 100$ FITs $\overline{P} = 390$ FITs

● **Point Estimate:**
  ✓ We demonstrated capability of 100 FITs.

● **Upper Bound Estimate:**
  ✓ 90% confidence bound is 390 FITs.

8.7 Influence of Acceleration Factors on Test Planning

Figure 8.28
Estimating accelerated testing time compression

- **An Acceleration Factor (AF) is used to estimate time compression**

$$\text{Test time} = \frac{Life\ time}{AF}$$

- **Acceleration Factors are estimated using experimental data or historical models**

Figure 8.26 actually illustrates the need for accelerated testing in order to be able to demonstrate that devices are capable of meeting a 100-FIT reliability objective with a reasonable experimental time frame and sample size. Time compression in accelerated testing is commonly achieved by using environmental stresses during testing. If a reasonable time compression model is known, an overall test acceleration factor can be estimated. In Figure 8.26, an acceleration factor of 100 was used, and test time was reduced from 100,000 hours (~ 11 years) to 1,000 hours (~ 1 month). This assumes that time is linearly compressed in accelerated testing as shown in Figure 8.28

Acceleration factors are based on historical models. The next chapter describes a number of models that are commonly used in accelerated testing. Such models are necessary to estimate the effects of raising the level of the appropriate stress to accelerate a potential device failure mode and effectively compress time. Thus, estimating time compression strongly influences test planning. Once the overall acceleration factor is estimated, tests can be properly planned. The acceleration factor can influence either the test time or the number of components needed in the test or both. For example, instead of testing 100 devices for 1,000 hours as shown in Figure 8.26, alternatively 1,000 devices could be tested for 100 hours. The important factor is device-hours in the test plan. This is illustrated in Figure 8.26 and Example 8.11 (also see Chapter 9, Example 9.7, showing the effects of time compression on test planning. Note that this is a simplified overview. Other statistical models and data handling techniques may apply when performing accelerated test planning and/or analysis (see Reference 8).

References

1. Klinger, D. J., Nakada, Y., and Menendez, M. A., *AT&T Reliability Manual,* Van Nostrand Reinhold, New York, 1990.
2. O'Connor, P. D. T., *Practical Reliability Engineering,* Wiley, New York, 1992.
3. Lewis, E. E., *Introduction to Reliability Engineering,* Wiley, New York, 1994.
4. Kennedy, J. B., and Neville, A. M., *Basic Statistical Methods for Engineers and Scientists,* Harper & Row, New York, 1986.
5. Kapur, K. C., and Lamberson, L. R., *Reliability in Engineering Design,* Wiley, New York, 1977.
6. Kececioglu, D. K., *Reliability Engineering Handbook,* Volumes 1 and 2, PTR Prentice Hall, Englewood Cliffs, NJ, 1991.
7. Nelson, W., *Applied Life Data Analysis,* Wiley, New York, 1982.
8. Nelson, W., *Accelerated Testing,* Wiley, New York, 1990.
9. Walpole, R. E., *Introduction to Statistics,* Macmillan, New York, 1982.
10. Kaplan, L. E., and Meier, P., "Nonparametric Estimation from Incomplete Observations," *Journal of the Acoustical Society of America,* Vol. 53, 282, pp. 457-481 (1958).

APPENDIX A

AT&T and Common Weibull Model Comparisons

In this Appendix, a comparison is provided for convenience of the AT&T Weibull reliability model and the common Weibull model provided in the literature. The comparison is given in Table 8.A.1. Table 8.A.2 provides conversions between models.

Table 8.A.1
Comparison between AT&T and common Weibull model

AT&T Weibull Model	Common Weibull Model
Reliability Function	
$R(t) = Exp(-\frac{\lambda_1}{1-y}t^{1-y})$	$R(t) = Exp[-(\frac{t}{\alpha})^{\beta}]$
Cumulative Distribution Function	
$F(t) = 1 - Exp(-\frac{\lambda_1}{1-y}t^{1-y})$	$F(t) = 1 - Exp\left[-(\frac{t}{\alpha})^{\beta}\right]$
Probability Density Function	
$f(t) = \lambda_1 t^y Exp(-\frac{\lambda_1}{1-y}t^{1-y})$	$f(t) = \frac{\beta}{\alpha^{\beta}}t^{\beta-1}Exp\left[-(\frac{t}{\alpha})^{\beta}\right]$
Hazard Rate (Instantaneous Failure Rate)	
$\lambda(t) = \lambda_1 t^{-y}$	$\lambda(t) = \frac{\beta}{\alpha}\left(\frac{t}{\alpha}\right)^{\beta-1}$
Cumulative Failure Rate (When $F(t)$ is < 0.1)	
$\lambda_{Cum}(t) = \frac{F(t)}{t} \approx \frac{\lambda_1}{1-y}t^{-y} = \frac{\lambda_{Inst.}(t)}{1-y}$	$\lambda_{Cum}(t) = \frac{F(t)}{t} \approx \frac{1}{\alpha}\left(\frac{t}{\alpha}\right)^{\beta-1} = \frac{\lambda_{Inst.}(t)}{\beta}$

Table 8.A.2
Conversions between AT&T and common Weibull model

AT&T Weibull Parameters α_{ATT}, λ_1	Common Weibull Parameters Shape Parameter β, Characteristic Life y
$\alpha_{ATT} = 1 - \beta$	$\beta = 1 - y$
$\lambda_1 = \frac{\beta}{\alpha_w^{\beta}}$	$\alpha = \left(\frac{1-y}{\lambda_1}\right)^{\frac{1}{1-y}}$

APPENDIX B

Helpful Microsoft® Excel Functions

Table 8.B.1
*Microsoft® Excel functions
for examples*

To aid the reader, Microsoft® Excel functions are provided in Table 8.B.1 to supplement this chapter's examples.

Chapter Example	Output Name	Excel Function	Excel Output
Example 8.4	Mean	= Average (2.2, 2.3, 2.5,...)	2.531
Example 8.4	Standard Deviation	= Stdev (2.2, 2.3, 2.5,...)	0.274
Example 8.5 Z (Area = 0.025)	Z-value	= Normsinv (1 – 0.025)	1.96
Example 8.6 Stand Dev = 2	Probability (2-sided)	= Normsdist (2) – (1 – Normsdist(2))	0.9545
Example 8.6 Stand Dev = 2	Probability (1-sided)	= Normsdist (2)	0.9773
Example 8.6 3.76, 4.06	PPM	= 1 – Normsdist (3.76) = 1 – Normsdist (4.06)	84.96E – 06 24.5E – 06
Example 8.8 N = 15 90% Confidence	t statistic	= TINV (1 – Conf.%/100, N – 1) = TINV (1 – 90/100, 15 – 1)	1.7613
Example 8.11 N = 100, Failures = 1 Single-Sided 90%	Chi square value	= Chiinv (1 – 0.9, 2*1 + 2) = Chiinv (1 – 0.9,4)	7.78

CHAPTER 9

Concepts in
Accelerated Testing

9.1 Introduction

The concept of accelerated testing is to compress time and accelerate the failure mechanisms in a reasonable test period so that product reliability can be assessed. The only way to accelerate time is to stress potential failure modes. These include electrical and mechanical failures. Figure 9.1 shows the concept of stress testing. Failure occurs when the stress exceeds the product's strength. In a product's population, the strength is generally distributed and usually degrades over time. Applying stress simply simulates aging. Increasing stress increases the unreliability (shown in Figure 9.1 as the overlap area between the strength and stress distributions) and improves the chances for failure occurring in a shorter period of time.

This also means that a smaller sample population of devices can be tested with an increased probability of finding failure. Stress testing amplifies unreliability so failure can be detected sooner. Accelerated life tests are also used extensively to help make predictions. Predictions can be limited when testing small sample sizes. Predictions can be erroneously based on the assumption that life-test results are representative of the entire population. Therefore, it can be difficult to design an efficient experiment that yields enough failures so that the measures of uncertainty in the predictions are not too large. Stresses can also be unrealistic. Fortunately, it is generally rare for an increased stress to cause anomalous failures, especially if common sense guidelines are observed.

9.2 Common Sense Guidelines for Preventing Anomalous Accelerated Testing Failures

Anomalous testing failures can occur when testing pushes the limits of the material out of the region of the intended design capability. The natural question to ask is: What should the guidelines be for designing proper accelerated tests and evaluating failures? The answer is: Judgment is required by management and engineering staff to make the correct decisions in this regard. To aid such decisions, the following guidelines are provided:

1. Always refer to the literature to see what has been done in the area of accelerated testing.

2. Avoid accelerated stresses that cause "nonlinearities," unless such stresses are plausible in product-use conditions. Anomalous failures occur when accelerated stress causes "nonlinearities" in the product. For example, material changing phases from solid to liquid, as in a chemical "nonlinear" phase transition (e.g., solder melting, intermetallic changes, etc.); an electric spark

Figure 9.1
Principal of accelerated testing

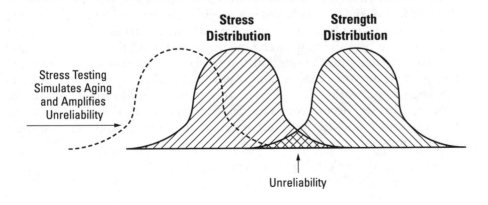

Stress Testing Simulates Aging and Amplifies Unreliability

Stress Distribution

Strength Distribution

Unreliability

compared to

avoiding high stresses or by allowing
or may not cause nonlinear stresses. In the latter test
design, a concurrent engineering design team reviews all failures and
decides if a failure is anomalous or not. Then a decision is made whether or
not to fix the problem. Conservative decisions may result in fixing some
anomalous failures. This is not a concern when time and money permit fixing all problems. The problem occurs when normal failures are labeled
incorrectly as anomalous and no corrective action is taken.

9.3 Time Acceleration Factor

The acceleration factor (A) is defined mathematically by Equation 9.1
where t is the typical life of a failure mode under normal use conditions and
t' is the life at accelerated test conditions:

$$A = \frac{t}{t'} \tag{9.1}$$

Since accelerated testing is designed to create failures in a shorter time frame,
the life under normal use conditions is usually much longer than the life under
accelerated test conditions, and A is much greater than 1. For example, an acceleration factor of 100 indicates that 1 hour in an accelerated stress environment
is equal to 100 hours in the normal use stress environment. Acceleration factors,
as denoted here, describe time compression. Acceleration factors may also be
put in terms of parameter change. The most common application is for estimating test time-compression using the time acceleration factor.

Acceleration factors are often modeled. For example, many failure modes
affected by temperature, such as chemical processes and diffusion, have what
is known as an Arrhenius reaction rate given by

$$Rate = B \exp\left\{\frac{-E_a}{K_B T}\right\} \tag{9.2}$$

where

B = a constant that characterizes the product failure mechanism
and test conditions (see Reference 1),
E_a = the activation energy in electron-volts (eV) of the failure mode,
T = the temperature (in degrees Kelvin), and
K_B = Boltzmann's constant (8.6173×10^{-5} eV/°K).

This is a thermodynamic expression that, while treated macroscopically to
describe failure kinetics, is obeyed in the microscopic world where elementary
reactions are taking place in accordance with the Arrhenius model. Particles
have a certain probability to overcome the potential barrier of height E_a and
become activated into the reaction taking place. As more and more elementary particles are consumed, a catastrophic event takes place at some point in
the macroscopic world. The rate is assumed to be inversely proportional to the
time that this will occur. For example, if an experiment is performed at two
temperatures T_1 and T_2, the failure times are then related to the rates at these
temperatures as

$$\frac{t_2}{t_1} = \frac{Rate(T_1)}{Rate(T_2)} \tag{9.3}$$

Combining equations 9.1, 9.2, and 9.3 yields the temperature acceleration factor

$$A_T = \frac{t_2}{t_1} = \exp\left\{\frac{E_a}{K_B}\left[\frac{1}{T_2} - \frac{1}{T_1}\right]\right\} \qquad (9.4)$$

The full model is shown in Figure 9.2 (Section 9.5). In order to evaluate the acceleration factor, the parameter activation energy E_a must be known or assumed for a particular failure mode. Often, historical information provides typical values for E_a, or these may be obtained through experimentation (see Example 9.2).

9.4 Applications To Accelerated Testing

To estimate test time compression and devise test plans that include sample size requirements, both acceleration models and statistical analysis are required (see Example 9.7). In this section, an overview of accelerated testing is provided in which potential failure mechanisms and acceleration models found in the literature are discussed.

Accelerated verification tests in microelectronics are designed to stress four types of failure mechanisms/modes. They are 1) thermomechanical mechanisms (e.g., package cracking, ohmic contacts, wire bond/lead integrity, thermal expansion mismatch problems, metal fatigue, creep, etc.), 2) nonmoisture-related thermochemical mechanisms (e.g., metal interdiffusion, intermetallic growth problems such as Kirkendall voiding, electromigration, MOS gate wearout, etc.), 3) moisture-related thermochemical mechanisms (e.g., surface charge effects, ionic leakage effects, dendrite growth, lead corrosion, galvanic corrosion, etc.), and 4) mechanical mechanisms (e.g., mechanical attachments, package integrity, fatigue, etc.). Combinations of these accelerated tests are required to properly stress each failure mechanism. The most common tests are Temperature Cycle, High-Temperature Operating Life (HTOL), Temperature-Humidity-Bias (THB), and Vibration, which are described here. Additionally, electromigration testing is described in this chapter. Temperature cycle stresses thermomechanical mechanisms; HTOL stresses nonmoisture-related thermo-chemical mechanisms; THB stresses moisture-related thermochemical mechanisms; and Vibration stresses mechanical failure mechanisms.

Additionally, many devices during manufacture receive some manufacturing stress. For example, Surface-Mount-Technology (SMT) devices are subject to solder-reflow processes. Therefore, to provide a realistic verification test procedure prior to accelerated reliability testing, devices should receive a pre-conditioning to simulate these stresses. In the case of SMT devices, a solder-reflow-type preconditioning test, such as described in JEDEC specification JESD22-A113, is commonly used.

9.5 High-Temperature Operating Life Acceleration Model

In High-Temperature Operating Life testing, devices are subjected to elevated temperature under bias for an extended period of time. Often, it is assumed that the dominant thermally accelerated failure mechanisms will follow the classical Arrhenius relationship (previously discussed). The traditional HTOL Arrhenius acceleration model is provided in Figure 9.2. The Arrhenius function is important. It is not only used in reliability to model temperature-dependent failure-rate mechanisms, but it also expresses a number of different physical thermodynamic phenomena (see Chapter 14). In

Figure 9.2

HTOL Arrhenius acceleration and linearized time-to-failure models

$$A_T = Exp\left\{\frac{E_a}{K_B}\left[\frac{1}{T_{Use}} - \frac{1}{T_{Stress}}\right]\right\}$$

$$Ln(t_f) = C + \frac{E_a}{K_B T}$$

Notation

A_T	=	temperature acceleration factor
T_{Stress}	=	test temperature (°K)
T_{Use}	=	use temperature (°K)
E_a	=	activation energy
K_B	=	8.6173×10^{-5} eV/°K (Boltzmann's constant)
t_f	=	time to failure
C	=	constant

Equation 9.2, we see that this factor is exponentially related to the activation energy. As the name connotes, in the failure process there must be enough thermal energy to be activated and surmount the potential barrier height of value E_a. As the temperature increases, it is easier to surmount this barrier and increase the probability of failure in a shorter time period. Thus, the activation energy parameter expresses a characteristic value that can be related to thermally activated failure processes. Each failure process has associated with it a barrier height E_a. In practice, when trying to estimate acceleration factor without knowing this value for each potential failure mechanism, a conservative value is used. For example, 0.7 eV is typically used for IC failure mechanisms and appears to be somewhat of an industry standard for conservatively estimating test times (see Examples 9.1 and 9.7). That is, a low value will over-estimate the test times and/or sample sizes needed to meet test objectives.

Obviously, the other important considerations are the actual use and stress temperatures. These estimates may also have errors. For example, to accurately assess time compression in testing, a device's junction temperature rise under bias needs to be taken out. This is illustrated in the next example.

▼ **Example 9.1** *Using the HTOL model*

Problem:

Estimate the test time to simulate 10 years of life in an HTOL test. The activation energies for the potential failure modes are unknown. Therefore, assume a conservative value of 0.7 eV for the activation energy. The device junction temperature rise is measured to be 15°C above ambient. The test temperature is 110°C, and the nominal use temperature is 40°C.

Solution:

Since the junction temperature rise is 15°C, then the actual use and test temperatures are

T_{Use} = 15°C + 40°C = 55°C
T_{Stress} = 15°C + 110°C = 125°C

From Figure 9.2, the acceleration factor is

$A_T = Exp$ {(0.7 eV/8.6173 × 10⁻⁵ eV/°K) × [1/(273.15 + 55) − 1/(273.15 + 125) °K]} = 77.6

From Equation 9.1, the test time to simulate 10 years of life (87,600 hours) is

Test Time = Life Time/A_T = 87600/77.6 = 1,129 hours

9.5.1 Estimating Activation Energy

Tests are often performed to determine a failure mechanism's activation energy. In this case, devices are separately tested in at least two different temperatures. Ideally, three or more temperatures can be used, then test results can be plotted on a semilog graph, and the data are fitted using a least-squares method. An example is the process reliability study shown in Figure 6.8 where a semilog plot is related to the linearized model in Figure 9.2. That is, if we plot the Mean-Time-To-Failure (MTTF) on the semilog axis versus *1/T*, then according to the equation

$$Ln(MTTF) = Const + \frac{E_a}{K_B}\left\{\frac{1}{T}\right\} \qquad (9.5)$$

the slope is E_a/K_B, and the activation energy can be determined as illustrated in the next example.

▼ **Example 9.2** *Determining the activation energy*

Problem:

The MTTFs at 250°C and 200°C are 731 and 10,400 hours, respectively, in Figure 6.8. Show that the activation energy is 1.13 eV and that the MTTF at 125°C is 1.95×10^{-6} hours as indicated in the figure.

Solution:

Equation 9.4 can be solved for E_a.
Then the activation energy is

$$E_a = K_B \frac{Ln\{MTTF_2 / MTTF_1\}}{(1/T_2 - 1/T_1)} \qquad (9.6)$$

Next, the acceleration factor at 125°C must be determined. Using the procedure in Example 9.1, we have

$$E_a = 8.6173\text{x}10^{-5}\ eV/°K\ \frac{Ln\,[10400/731]}{[1/(273.16+200)-1/(273.16+250)]\ °K} = 1.133eV$$

T_{Use} = 125°C
T_{Stress} = 200°C
A_T = Exp $\{(1.133\ eV/8.6173 \times 10^{-5}\ eV/°K) \times [1/(273.15 + 125) - 1/(273.15 + 200)\ °K]\} = 187.6$

From Equation 9.1, the MTTF (at 125°C) = MTTF (at 200°C) × $A_T = 10400 \times 187.7 = 1.951 \times 10^6$ hours. The answer is a bit off of the value shown in Fig. 6.8 due to round-off error.

9.6 Temperature-Humidity-Bias Acceleration Model

In THB, test devices are put at elevated temperatures and humidity under bias for an extended period of time. For example, the most common THB test is a 1000-hour test at 85°C and 85 percent Relative Humidity. One of the most common THB models used in the industry is a 1989 Peck model (see Reference 3) shown in Figure 9.3. A derivation is provided in Chapter 14, Section 14.5.2. This includes a relationship between life-and-temperature (Arrhenius model) and life-and-humidity (Peck model), so that the product of the two separable factors yields an overall acceleration factor.

$$A_T = Exp\left\{\frac{E_a}{K_B}\left[\frac{1}{T_{Use}} - \frac{1}{T_{Stress}}\right]\right\}$$

$$A_H = \left(\frac{R_{Stress}}{R_{Use}}\right)^m$$

$$A_{TH} = A_T A_H$$

$$Ln(t_f) = C + \frac{E_a}{K_B T} - m\ Ln(R)$$

Notation

A_H	=	humidity acceleration factor
A_T	=	temperature acceleration factor
A_{TH}	=	temperature-humidity acceleration factor
R_{Stress}	=	relative humidity of test
R_{Use}	=	nominal use relative humidity
T_{Stress}	=	test temperature
T_{Use}	=	nominal use temperature
m	=	humidity constant
E_a	=	activation energy
t_f	=	time to fail
C	=	constant

Figure 9.3

THB Peck acceleration and linearized time to failure models

▼ **Example 9.3** *Using the THB model*

Problem:

If a THB test is performed at 85%RH and 85°C, what is the acceleration factor relative to a 40%RH and 25°C environment, assuming an activation energy of 0.7 eV and a humidity constant of 2.66? How many test hours are required to simulate 10 years of life? How many test hours are required in a HAST chamber (see Chapter 5) to simulate 10 years of life at 85%RH and 110°C?

Solution:

The temperature acceleration factor is
$A_T = $ Exp {(0.7 eV/8.6173 × 10^{-5} eV/°K) × [1/(273.15 + 25) − 1/(273.15 + 85)°K]} = 96

The humidity acceleration factor is
$A_H = $ (85%RH/40%RH)$^{2.66}$ = 7.43

Therefore, the combined temperature humidity acceleration factor is
$A_{TH} = $ 96 × 7.43 = 713

The simulated test time to equate this to 10 years (87,600 hours) is
Test time = (87,600 hours/713) = 123 hours

The temperature acceleration factor for the HAST test is
$A_T = $ Exp {(0.7 eV/8.6173 × 10^{-5} eV/°K) × ([1/(273.15 + 25) − 1/(273.15 + 110)°K]} = 421.8

The humidity acceleration factor is the same as in the first part of the problem so that
$A_{TH} = $ 421.8 × 7.43 = 3132.2

The simulated test time to equate this HAST test to 10 years is
HAST test time = (87600 hours/3132) = 28 hours

When Peck originally proposed this model, he reviewed all published life-in-humidity conditions versus life at 85°C/85%RH for epoxy packages. His results found good agreement with the model. Fitted data found nominal values for E_a to lie between 0.77 and 0.81 and nominal values between 2.5 and 3.0 for m. A thorough study by Texas Instruments (see Reference 4) on PEM moisture-life monitoring found the activation energy values up around 0.9 eV. Such trends in the literature indicate higher-activation energies, which correspond to trends in improved semiconductor reliability.

9.7 Temperature Cycle Acceleration Model

In Temperature Cycle, test devices are subjected to a number of cycles of alternate high and low temperature extremes. This cyclic stress produced in temperature cycling is related to thermal expansion and contraction undergone in the material. To relate field usage to accelerated test conditions, the most widely used model in industry is the Coffin-Manson (see Reference 1) model. This is a simple model used for estimating the temperature cycle acceleration factor (see Figure 9.4). A derivation of this model is provided in Chapter 14, Section 14.4.2.

Reasonably estimating the acceleration factor depends on the failures being caused by fatigue subject to the Coffin-Manson law for cyclic strain versus the number of cycles to failure. Values between 2 to 4 have typically been reported in the literature for K. These values are related to the specific design. A value of 2.5 is commonly used for solder-joint fatigue, while 4 is often reported for IC interconnection failures. The lower value (2.5) is a good value for conservative estimates.

▼ **Example 9.4** *Using the Temperature Cycle model*

Problem:
Estimate the number of test temperature cycles to simulate 10 years of life in the field for a test that cycles between −55°C and 150°C. It is estimated that field conditions cycle nominally between −5°C to 25°C twice a day. Assume a conservative temperature cycle exponent of 2.5.

Figure 9.4
Temperature Cycle acceleration and linearized cycle to failure models

$$A_{TC} = \frac{N_{Use}}{N_{Stress}} = \left(\frac{\Delta T_{Stress}}{\Delta T_{Use}} \right)^K$$

$$Ln(N_f) = C - K\ Ln(\Delta T)$$

Notation
A_{TC} = temperature cycle acceleration factor
N_{Stress} = number of cycles tested
N_{Use} = equivalent number of field cycles
ΔT_{Stress} = temperature cycle test range
ΔT_{Use} = nominal daily temperature change in the field
K = temperature cycle exponent
N_f = number of cycles to failure
C = constant

Solution:
First, use the expression in Figure 9.4 to find the temperature cycle acceleration factor as
$A_{TC} = (\Delta T_{Stress}/\Delta T_{Use})^K = (205°C/30°C)^{2.5} = 122.$
In 10 years, the device will be cycled $2 \times 365 \times 10 = 7300$ cycles. Therefore, from Figure 9.4, the number of test cycles to simulate this is
$N_{Stress} = N_{Use}/A_{TC} = 7300/122 = 60$ cycles

9.8 Vibration Acceleration Model

In Vibration, devices are mounted on a dynamic shaker table and subject to either a random or sinusoidal-type vibration profile. Common random vibration tests are most frequently specified in terms of Power Spectral Density (PSD) levels (see Figure 9.5). Figure 9.5 illustrates the possible PSD test profile levels related to a similar particular use environment. The PSD function describes the distribution of vibration energy with respect to frequency. The amount of time compression that can be accomplished is related to the PSD test level and use level. Estimates of time compression can be made once the use level estimate and spectral density profile are established. The traditional classical time-compression model (MIL-STD 810E) is a power law model (see Figure 9.6). A derivation of this model is provided in Chapter 14, Section 14.4.3.

In applying this model, it is important to understand the failure mechanism since, in a random vibration-loading environment, the resonance of the material can dominate the fatigue life. Here, maximum vibration amplitudes and stress occur. However, fatigue failures are not always dominated by the fundamental resonance mode. In practice, many of the stress peaks in the use environment may fall below the fatigue limit of the material, while others will be above the fatigue limit. It follows that an accelerated life test based upon this model should inherently be conservative. However, because most of the fatigue damage occurs at the highest stress peaks in both the test and in actual use, the degree of conservatism is not excessive. As noted in Figure 9.5, the PSD is in units of G^2/Hz. The square of the G stress level at the resonance frequency is directly proportional to the PSD level ($W \sim G^2$), so the model can be put in terms of either random or sinusoidal-resonance G-stress loading. In the model, the fatigue parameter is related to experimental slope of the stress to cycles to failure data. This exponent varies depending on the fatigue life of the materials involved. For example, the value of $b \approx 5$ is commonly used for electronic boards. However, conservative value for the fatigue parameter b is about 8 (e.g., $M_b = 4$). MIL STD-810E (514.4-46) recommends $b = 8$ for random loading.

Figure 9.5

Example of common PSD test levels

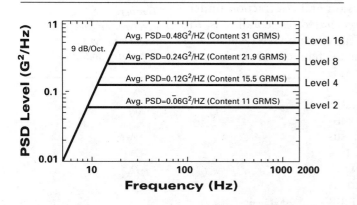

▼ **Example 9.5** *Using the Vibration model*

Problem:

Estimate the test time to simulate 10 years of life in the field for an assembly that is tested to a Level 4 PSD random vibration test condition, shown in Figure 9.5. It is estimated that the assembly will undergo a worst-case Level 1-type random vibration exposure 1 percent of the assembly's life. The rest of the assembly's life is relatively benign in terms of vibration exposure.

Figure 9.6

Vibration acceleration and linearized time to failure models

$$A_V = \frac{T_{Use}}{T_{Stress}} = \left(\frac{W_{Stress}}{W_{Use}}\right)^{M_b}$$

$$\left(\frac{W_{Stress}}{W_{Use}}\right)^{M_b} = \left(\frac{Gf_{Stress}}{Gf_{Use}}\right)^{2M_b}$$

$$Ln(t_f) = C - M_b\ Ln(W)$$

Notation

A_V	=	vibration acceleration factor
T_{Stress}	=	vibration duration
T_{Use}	=	vibration duration (nominal)
W	=	random vibration input PSD across the resonance bandwidth (G^2/Hz) W_{Stress} is the PSD test stress and W_{Use} nominal use PSD
Gf	=	resonant G sinusoid vibration level
M_b	=	b/2 where b is the fatigue parameter
t_f	=	time to failure
C	=	constant

Solution:

First, use the expression in Figure 9.6 to find the vibration acceleration factor. Since Level 4 has a PSD of 0.12 G^2/Hz, then Level 1 is 0.03 G^2/Hz. Therefore,

$A_V = (W_{Stress}/W_{Use})^{Mb} = (0.12/0.03)^4 = 256.$

In 10 years, the device will be exposed to a Level 1 vibration for about $87,600 \times 0.01 = 876$ hours. Therefore, from Figure 9.6, the number of test cycles to simulate this is

$T_{Stress} = T_{Use}/A_V = 876/256 = 3.5$ hours.

9.9 Electromigration Acceleration Model

Electromigration is a failure mechanism caused in a microelectronic conductor exposed to high current densities or a combination of high temperature and current density. The most common failure mode is a conductor open. This failure mechanism comes about from high current densities that create crowded electron flux in the microelectronic conductive path. Often, the term "electron wind" has been historically used for the scattering mechanism causing failure. The metal reaches a stage at which collision between the electrons and film atoms and defects sites becomes catastrophic. Electron scattering from defect sites is considered to dominate. The collision rate increases to the point that atoms of the metal film drift in the direction of the electron flow. Eventual catastrophic problems result due to local inhomogenous regions in the metal combined with the metal movement.

Figure 9.7
Electromigration acceleration and linearized time to failure model

Generally, the Black equation (see References 6 and 7) is widely used for making MTTF electromigration predictions in the literature. For the electromigration acceleration factor due to the Black equation, see Figure 9.7. Numerous values for the Black equation parameters n and E_a have been reported in the literature. As lower values are used, the estimates become more conservative. Numerous experiments have been performed under various stress conditions in the literature, and values for n have been reported in the range between 2 and 3.3 and between 0.5 to 1.1 eV for E_a.

$$A_J = \left(\frac{J_{Stress}}{J_{Use}}\right)^n Exp\left\{\frac{E_a}{K_B}\left[\frac{1}{T_{Use}} - \frac{1}{T_{Stress}}\right]\right\}$$

$$Ln(t_f) = C + \frac{E_a}{K_B T} - nLn(J)$$

Notation

A_J = electromigration acceleration factor
T_{Stress} = test temperature (°K)
T_{Use} = use temperature (°K)
E_a = activation energy
K_B = 8.6173×10^{-5} eV/°K (Boltzmann's constant)
J = current density
n = current density exponent
t_f = time to failure
C = constant

▼ **Example 9.6** *Using the Electromigration model*

Problem:
An electromigration experiment performed on aluminum conductors at 185°C and a current density of 3×10^5 A/cm² found an MTTF of 2000 hours. Estimate the MTTF at a use condition of 100°C and a current density of 2×10^5 A/cm². Use conservative parameter estimates of $E_a = 0.5$ eV and $n = 2.0$.

Solution:
First, find the temperature acceleration factor, which is
$A_T = Exp\{(0.5 \text{ eV}/8.6173 \times 10^{-5} \text{ eV/°K}) \times [1/(273.6 + 100) - 1/(273.6 + 185) \text{ °K}]\} = 17.9$

The current density factor is
$A_c = (3 \times 10^5 \text{ A/cm}^2 /2 \times 10^5 \text{ A/cm}^2)^2 = 2.25$

The product provides the electromigration acceleration factor
$A_J = A_T A_c = 17.9 \times 2.25 = 40.3$

The MTTF at use condition can then be estimated as
$MTTF_{Use} = MTTF_{Stress} \times A_J = 2000 \times 40.3 = 80,600$ hours = 9.2 years

9.10 Failure-Free Accelerated Test Planning

There are numerous types of accelerated tests. Any test that in some way accelerates environmental use conditions is an accelerated test. Two of the most common types of accelerated tests used in industry are catastrophic and failure-free testing. In a catastrophic accelerated test, a frequent objective is to estimate the failure rate at a use condition. A number of examples to estimate the MTTF at use condition (see Example 8.7) have been provided. Note that in each case, one had to assume conservative values of model parameters such as the activation energy. Example 9.2 illustrated how, in a process reliability study, the activation energy for a particular failure mode can be estimated.

In Chapter 4, Design Maturity Testing (DMT) was discussed. DMT is based on failure-free testing. The main objective of a DMT test is to determine whether a design will meet its reliability objective at a certain level of confidence. This requires that a statistically significant sample size be tested in a number of different stress tests. This topic was introduced in Section 4.6. In Chapter 8, an example was provided on accelerated demonstration versus statistical sample planning. However, at this point, we would like to illustrate how to conservatively plan a DMT to demonstrate that a particular reliability objective can be met.

▼ **Example 9.7** *Designing a Failure-Free Accelerated Test*

Problem:
Plan accelerated tests for a failure-free DMT to demonstrate that a plastic-packaged IC will meet its reliability objective of 400 FITs (Objective 4, Figure 4.3) at the 90 percent confidence level. Estimate the sample size required and test times needed to show that this component is failure-free of any HTOL, THB, and TC type failure modes. Use the acceleration factors found in Examples 9.1, 9.3, and 9.4 in your design.

Solution:
A full DMT test for this component will include nonaccelerated tests as well. Figure 4.5 illustrates the concept, and Chapter 4 describes DMT in detail. To design the accelerated testing portion, first estimate a practical test duration. For example, we can target the test to last about a 1000 hours for HTOL and THB, and about 100 temperature cycles. Once we have fixed the test time, we next must estimate a statistically significant sample size at the 90 percent confidence level. We can assume that each test will check for different failure modes. This means that each test is allocated a portion of the failure rate. One allocation plan is described in Section 4.2, where THB-, TC-, and HTOL-type failure modes were assigned 20 percent, 30 percent, and 50 percent, respectively, of the total reliability. Using this plan, the 400 FITs are broken up with 80, 120, and 200 FITs to THB, TC, and HTOL tests, respectively. At this point, a single-sided chi-square estimate for sample-size planning can be used. This is detailed in Section 4.6 where the sample size N is given

$$N(HTOL) = \chi^2(90\%, 2Y + 2)/2\lambda At$$

For example, the TC values are
$Y = 0$ Failures
$\chi^2(90\%, 2) = 4.605$
$\lambda = 120$ FITs $= 1.2 \times 10^{-7}$ failure/hour
$A = 122$ (from Example 9.4)
$t = 100$ cycles $\times$ 24 Hours $= 2400$ equivalent test hours

Thus,
$$N = 4.605/(2 \times 1.2 \times 10^{-7} \times 122 \times 2400) = 66 \text{ devices}$$

Using this same approach for the other tests, the results are summarized in Table 9.1.

Table 9.1
Summary of DMT for Example 9.8

Accelerated Test	Acceleration Factor	Test Time	FITs	Sample Size
HTOL	78	1000	200	148
THB	713	1000	80	41
TC (100 Cycles)	122	2400	120	66

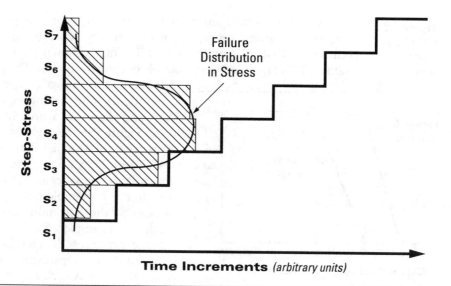

9.11 Step-Stress Testing

Figure 9.8
Concept of Step-Stress Testing

Step-Stress Testing is an alternative test to life testing. In Step-Stress Testing, usually a small sample of devices is exposed to a series of successively higher and higher steps of stress. At the end of each stress level, measurements are made to assess the results to the device. The measurements could be simply to assess if a catastrophic failure has occurred or to measure the resulting parameter shift due to the step's stress. Constant time periods are commonly used for each step-stress period. This provides for simpler data analysis. The concept is shown in Figure 9.8. Note that the failure distribution over the stress levels is usually experimentally found to be normally distributed. This is a consequence of a normally distributed strength distribution (see Figure 9.1). Therefore, the plot of CDF versus stress should be plotted on a normal probability plot. In general, if the data does not fit a normal probability plot, other distributions should be tried. An example of a CDF that is normally distributed over temperature step-stresses is provided in Example 9.10 in Section 9.12.1 (see Section 8.4.3 on normal probability plotting). Although not shown in this plot, stress data at high stress levels will often deviate from normality. This most likely indicates that high stress levels can cause nonlinear changes in the material under test such as a phase change; thus the materials strength departs from observed normality.

There are a number of reasons for performing a step-stress test, including:

- Aging information can be obtained in a relatively short period of time. Common step-stress tests take about 1 to 2 weeks, depending on the objective.
- Step-stress tests establish a baseline for future tests. For example, if a process changes, quick comparisons can be made between the old process and the new process. Accuracy can be enhanced when parametric change can be used as a measure for comparison. Otherwise, catastrophic information is used.
- Failure mechanisms and design weaknesses can be identified along with material limitations. Failure-mode information can provide opportunities for reliability growth. Fixes can then be put back on test and compared to previous test results to assess fix effectiveness.
- Data analysis can provide accurate information on the stress distribution in which the median-failure stress and stress standard deviation can be obtained. This then provides an MTTF estimate at the median failure stress level.

9.11.1 Temperature Step-Stress (TSS)

Probably the most common step-stress is temperature. In Temperature Step-

Figure 9.9
*Data plot from
two Temperature
Step-Stress Tests*

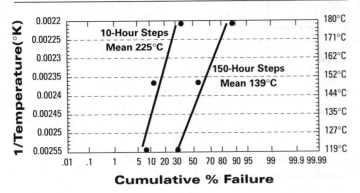

Stress, catastrophic data are plotted on a normal probability plot with the cumulative failure percent versus 1/temperature (in °K). Data are plotted this way because the CDF is a function of 1/T. This is shown later in Section 9.12.1, Example 9.10. Figure 9.9 is an example of this type of plot. The data (see Example 9.8) resulted from two Temperature Step-Stress experiments, one having equal 10-hour time steps and the other with 150-hour time steps. From Figure 9.9, the mean stress points (where 50 percent of the distribution has failed) are 139°C and 225°C. Since these are mean stress points, it also provides an MTTF estimate for the step time. For example, these points can be used to estimate the activation energy for the failure mode (see Example 9.8).

Since step-stress data has been carried out in constant time steps, some accumulation of residual effects at each step occurs from previous stress-steps. This makes the data slightly off. A step-stress correction can improve accuracy of estimating the mean stress point. The next example will illustrate how step-stress accuracy can be improved.

▼ **Example 9.8** *Temperature Step-Stress analysis*

Problem:

The data for two Temperature Step-Stress times from an experiment is provided in Table 9.2. Twenty-four parts were tested. In the first Temperature Step-Stress test, 10-hour steps were used; in the second experiment, 150-hour steps were used. Plot the data, determine the mean stress values, and estimate the activation energy for the two tests. Provide a correction for each data point and re-estimate the activation energy from the correction. When is it reasonable to provide a correction? What is the estimated MTTF at 25°C?

Solution:

The number of failures is shown in Table 9.2. In Temperature Step-Stress data, 1/T (°K) is plotted versus the cumulative percent failure. Therefore, data are arranged in the table for plotting directly. Note that the cumulative percent failure is obtained as described in Chapter 8 using i/n+1 values. The data has been plotted previously in Figure 9.9, and the mean stress values are 225°C and 139°C for the 10-hour and 150-hour tests, respectively. Note these times are MTTF values at their respective temperatures. With these values, an activation energy can be obtained similarly to Example 9.2 as

$$E_a = 8.6173 \times 10^{-5} \text{ eV/°K } ln[150/10]/\{1/(273.15 + 139) - 1/(273.15 + 225)\} = 0.557 \text{ eV}$$

The accuracy of this data can be improved with a Temperature Step-Stress

Table 9.2
*Temperature
Step-Stress data
for Figure 9.9*

Temperature (°C)	1/T(°K)	No. Failures 10-Hour Steps	Cumulative % Failure 10-Hour Steps	No. Failures 150-Hour Steps	Cumulative % Failure 150-Hour Steps i/(n + 1)
120	0.002545	2	8	8	32
150	0.002364	1	12	6	56
180	0.002208	6	36	8	88

Test Temperature (°C)	10-Hour Time Equivalent	150-Hour Time Equivalent	Temperature Correction (Same for both time steps!)
120	10	150	120
150	13.1	196.5	157.6
180	14.7	221.1	192.6

correction. If the stress-steps are incrementally large enough, usually a correction is not necessary. In this experiment, the stress steps are 30°C apart, which is borderline. Therefore, a correction may improve accuracy. Consider the 10-hour TSS data. First, correct the 150°C data point. Devices received 10 hours of exposure at 150°C, but they had already been exposed to 10 hours at 120°C. According to Example 9.1, the acceleration factor between 120°C and 150°C with an E_a of 0.56 is

$$A_T = Exp\{(0.56 \text{ eV}/8.6173 \times 10^{-5} \text{ eV}/°K) \times [1/(273.15 + 120) - 1/(273.15 + 150)] °K\} = 3.23$$

Table 9.3

Corrected Temperature Step-Stress data

Therefore, devices failing at the 150°C point had received 10 hours at 120°C and now 10 hours at 150°C prior to failing. The total exposure is actually equivalent to

10 + 10/3.23 = 13.1 hours

at 150°C. However, to replot this data point more accurately as a 10-hour failure point, find the temperature at 10 hours that is equivalent to 13.1 hours of exposure at 150°C. To do this, solve Equation 9.4 for T_2 in degrees centigrade. This is

$$T_2(°C) = [(0.000086173/E_a) \times ln(t_1/t_2) + 1/(T_1 + 273.15)]^{-1} - 273.15$$

Inserting the appropriate values, the temperature correction is

$$T_{Correction}(°C) = [(0.000086173/0.56) \times ln(10/13.1) + 1/(150°C + 273.15)]^{-1} - 273.15 = 157.6°C$$

Therefore, the corrected temperature is 157.6°C. This is a more accurate temperature value for plotting the failures at this 10-hour step-stress point. In a similar manner, one can estimate that the 10-hour equivalent temperature at 180°C is 192.6°C. The corrected values are shown in Table 9.3. As an exercise, the reader can verify these values. The data can now be replotted. This is not shown here, as the plot is very similar to Figure 9.9. However, the means obtained from the corrected plot are 224°C and 143°C for the 10-hour and 150-hour steps, respectively. With these new values, our estimates can be refined for the activation energy. The new estimate with these corrected temperatures is

$$E_a = 8.6173 \times 10^{-5} \text{ eV}/°K \ ln[150/10]/\{1/(273.15 + 143) - 1/(273.15 + 224)\} = 0.596 \text{ eV}$$

Using this value, the MTTF at 25°C can be predicted. The acceleration factor between 25°C and 143°C is 719. Since the MTTF at 143°C is 150 hours, then at 25°C the predicted MTTF is 107,813 hours (= 719 × 150).

It is important to note when it is reasonable to provide Temperature Step-Stress corrections. Since in many step-stress experiments, devices are measured only once (at the end of each step), the exact failure time is not known. In this case, it is probably not worth providing a correction, especially if the correction is relatively small, since devices could have failed at any point during the step time. However, if devices are monitored during the test period and exact failure times are recorded, then the correction can be helpful.

9.12 Describing Life Distributions as a Function of Stress

It is instructive to illustrate how to incorporate a stress model into a life distribution. This can be illustrated for both the power law form and the Arrhenius function. These will be incorporated into the CDF and PDF for the log-normal distribution. Consider the Arrhenius temperature and the vibration models given in Figures 9.2 and 9.6. The time to failure is written in linear form and repeated here for convenience. From Figure 9.2, this is

$$Ln(t_f) = C + \frac{E_a}{K_B T} \tag{9.7}$$

and from Figure 9.6, it is

$$Ln(t_f) = C - Mb\ Ln(W) \tag{9.8}$$

Experimentally, the time to failure can be assessed at any time. For the log-normal distribution, these parameters apply to the median time to failure, $t_f = t_{50}$. This allows for a direct substitution into the log-normal distribution functions of Figure 8.14. Inserting the Arrhenius function into the PDF reads

$$f(t,T) = \frac{1}{\sigma_t(T)t\sqrt{2\pi}}\ Exp\left\{-\frac{1}{2}\left(\frac{Ln(t)-\left(C+\frac{E_a}{K_B T}\right)}{\sigma_t(T)}\right)^2\right\} \tag{9.9}$$

and for the vibration model, this is

$$f(t,W) = \frac{1}{\sigma_t(W)t\sqrt{2\pi}}\ Exp\left\{-\frac{1}{2}\left(\frac{Ln(t)-\left(C-Mb\ Ln(W)\right)}{\sigma_t(T)}\right)^2\right\} \tag{9.10}$$

Similarly, inserting the Arrhenius model into the Cumulative Distribution Function (using the error function form in Figure 8.15) reads

$$F(t,T) = \frac{1}{2}\left[1 + erf\left(\frac{Ln(t)-\left(C+\frac{E_a}{K_B T}\right)}{\sqrt{2}\sigma_t(T)}\right)\right] \tag{9.11}$$

and for the vibration model, this is

$$F(t,W) = \frac{1}{2}\left[1 + erf\left(\frac{Ln(t)-\left(C-Mb\ Ln(W)\right)}{\sqrt{2}\sigma_t(T)}\right)\right] \tag{9.12}$$

Similar expressions can be found for the CDF and PDF of any life distribution function when t_f is appropriately found. As an exercise, find these for the common Weibull function given in Table A.2, Chapter 8. (Hint: Assume that $t_{.632} = t_f$; this then is the characteristic life α_w in the table.)

9.12.1 Stress-Dependent Standard Deviation

In these above expressions, note that sigma could be temperature-dependent. Usually, this is not the case. One model to determine this from life test data over temperature is

$$Ln(\bar{t}_f) = C + \frac{E_a}{K_B T} = C_{50\%} + \frac{E_{a50\%}}{K_B T} \tag{9.13}$$

Here the constants can be identified from fitting the Mean-Time-To-Failure fit over stress. Additionally, we could also obtain a fit to the data at the 16th percentile points in the distribution over stress denoted here as

$$Ln(t_f)_{16\%} = C_{16\%} + \frac{E_{a16\%}}{K_B T} \tag{9.14}$$

This gives a model for sigma based on the physical aging law and the data itself as

$$\sigma_t(T) = Ln(\bar{t}_f)_{50\%} - Ln(t_f)_{16\%} = \Delta C + \frac{\Delta E_a}{K_B T} \tag{9.15}$$

In this view according to (9.13) and (9.15), for sigma to depend on temperature, ΔE_a must be non zero. This indicates that E_a is distribution dependent. Our intuition would tell us this might occur with a bimodal failure mode, or as a result of statistical resolution difficulties. In the statistical case, we might attempt an alternate method to fit the data like the Weibull distribution. However, in the bimodal case, we are likely to find that the Weibull Beta slope would have a similar behavior on temperature as observed with a lognormal sigma.

▼ **Example 9.9** *CDF as a function of stress*

Problem:
For the vibration function, let $C = -7.82$, $Mb = 4$, and find $F(t,W)$ for $t = 10$ years and $W = 0.0082$ G^2/Hz. Find F at 10 years. Use $\sigma = 2.2$ for your estimate. If the stress level is reduced by a factor of 2, what is F?

Solution:
Inserting these values into the CDF above reads,

$$F(t,W) = \frac{1}{2}\left[1 + erf\left(\frac{\ln(87600) - \left(-7.82 - 4\ Ln(0.0082)\right)}{\sqrt{2}(2.2)}\right)\right] \tag{9.16}$$

or

$$F(87600, 0.0082) = \frac{1}{2}\left[1 + erf\left(\frac{-0.0139}{\sqrt{2}(2.2)}\right)\right] = \frac{1}{2}\left[1 - erf\left(\frac{0.0139}{\sqrt{2}(2.2)}\right)\right] = 0.497$$

Thus, at this stress level, 49.7 percent of the distribution is anticipated to have failed in 10 years. (Note: In the above derivation, the error function values can be found from tables or in Microsoft® Excel type, = erf(0.00447) to obtain the above value.) If the stress level is reduced by a factor of 2, then $W = 0.0041$ G2/Hz. The anticipated percent failure at 10 years is reduced to $F(87600, 0.0041) = 10.27\%$.

▼ **Example 9.10** *Relationship between a stress and time standard deviation*

Problem:

Provide a CDF model for the temperature stress distribution and find a relationship between the standard deviation for stress, σ_T, and for time, σ_t. Use this relationship to estimate α_t from a step-stress experiment. The temperature step-stress experiment was run with 24-hour increments. Data indicated that 50 percent of the devices fail at 250°C (523°K) and 16 percent fail at 200°C (473°K). The failure mechanism activation energy is 1.3 eV.

Solution:

The model for the combined CDF time and stress distribution is given above by $F(t,T)$. We can substitute in the general relationship for the time to fail

$$Ln(t_f) = C + \frac{E_a}{K_B T} \qquad (9.17)$$

which holds for both the MTTF and for any time giving

$$F(t,T) = \frac{1}{2}\left[1 + erf\left(\frac{\left(C + \frac{E_a}{K_B T}\right) - \left(C + \frac{E_a}{K_B \overline{T}}\right)}{\sqrt{2}\sigma_t}\right)\right] \qquad (9.18)$$

Simplifying this expression gives

$$F(t,T) = \frac{1}{2}\left[1 + erf\left(\frac{\left(\frac{1}{T} - \frac{1}{\overline{T}}\right)}{\sqrt{2}\frac{K_B}{E_a}\sigma_t}\right)\right] = \frac{1}{2}\left[1 + erf\left(\frac{\left(\frac{1}{T} - \frac{1}{\overline{T}}\right)}{\sqrt{2}\sigma_S}\right)\right] \qquad (9.19)$$

By comparison, the relationship between the standard deviations is

$$\sigma_t = \frac{E_a}{K_B}\sigma_S \qquad (9.20)$$

To solve for the second part of the problem, note from any normal distribution table that $(1/T)_{50\%} - (1/T)_{16\%}$ is approximately one standard deviation apart. Therefore,

$$\sigma_S \approx \frac{1}{473} - \frac{1}{523} = 0.000202$$

and

$$\sigma_t = \frac{1.3 eV}{8.62 x 10^{-5}} 0.000202 = 3.05$$

9.13 Summary

In this chapter, accelerated testing has been described. The general objective in accelerated testing is to accelerate time and predict information about the product's reliability. However, a further objective not discussed is to grow reliability through testing and fixing failure modes. This is the topic of the next chapter.

References

1. Nelson, W., *Accelerated Testing*, Wiley, New York, 1990.

2. Feinberg, A. A., "The Reliability Physics of Thermodynamic Aging," *Recent Advances in Life-Testing and Reliability*, edited by N. Balakrishnan, CRC Press, Boca Raton, FL.

3. Peck, D. S., "Comprehensive Model for Humidity Testing Correlation," *International Reliability Physics Symposium*, 1986, pp. 44-50.

4. Tam, "Demonstrated Reliability of Plastic-Encapsulated Microcircuits for Missile Applications," *IEEE Transactions on Reliability*, Vol. 44, No. 1, 1995, pp. 8-13.

5. Denson, W. K., "A Reliability Model for Plastic-Encapsulated Microcircuits," *Institute of Environmental Sciences Proceedings*, 42nd Annual Meeting, 1996, pp. 89-96.

6. Black, J. R., "Metallization Failures in Integrated Circuits," *Technical Report*, RADC-TR-68-43 (Oct. 1968).

7. Black, J. R., "Electromigration – A Brief Survey and Some Recent Results," *IEEE Transactions on Electron Devices*, Vol. ED-16, No. 4, 1969, p. 338.

CHAPTER 10

Accelerated Reliability Growth

10.1 Introduction

In this chapter, the focus is on Reliability Growth management and planning in a commercial environment. This is because Reliability Growth management/planning can be the single most important factor in achieving customer satisfaction. To do this effectively in a manufacturing environment with numerous product types, we must depart a bit from the traditional Reliability Growth methodology and broaden the concepts to encompass the accelerated stage gate processes.

Traditionally, the need for Reliability Growth planning has been for large subsystems or systems (see Reference 1). This is simply because of the greater risk in new product development at that level compared to the component level. Also, in programs where one wishes to push mature products or complex systems to new reliability milestones, inadequate strategies will be costly. A program manager must know if Reliability Growth can be achieved under required time and cost constraints. A plan of attack is required for each major subsystem so that system-level reliability goals can be met.

The approach here differs in that Reliability Growth planning is recommended for all new "platforms," whether they are complex subsystems or simple components. In a commercial environment with numerous product types, the emphasis must be on platforms rather than products. Often there may be little time to validate, let alone assess, reliability (see References 2 and 3). Yet, without some method of assessment, platforms could be jeopardized. Accelerated testing is, without question, the featured Reliability Growth tool for industry. It is important to devise reliability planning during development that incorporates the most time- and cost-effective testing techniques available. Plans can now take advantage of the advances in the area of accelerated testing, such as current test equipment like Highly Accelerated Stress Test (HAST), historical acceleration factors, system-level acceleration factor estimation procedures (see References 2–13), and so forth. Advanced tools are now available for improving products and process, such as:

Figure 10.1
Concept of Accelerated Reliability Growth

- Design of experiments
- Taguchi methods
- Multivariable analysis
- Thermal analysis
- Parametric reliability analysis

✓ The only way to improve (grow) reliability is to find and fix failure modes.

✓ The only way to find a hidden failure mode is to stress it.

✓ Raising the level of the appropriate stress is the only way to accelerate this process.

Failure Modes and Effects Analysis (FMEA) is also viewed as a Reliability Growth planning tool, as it is best to grow reliability as early as possible. These advances fit well into the stage gate process with the appropriate application of classical Reliability Growth originally described in Military Handbook-189 (see Reference 1). In the broadest sense, however, the concept of Accelerated Reliability Growth (ARG) testing, as outlined in Figure 10.1, is simple and is based on finding and fixing failure modes.

Today's competitive market requires thorough planning, especially since platform complexity has increased dramatically as competition and technological advances have driven down size and costs. It would be unwise to develop an expensive platform without planning for "platform Reliability Growth." Properly applied, Reliability Growth methods are powerful. Traditional methodology provides us with many valuable tools, such as test planning, growth tracking and assessment, fix-effectiveness factor estimation, corrective action review team operations, and Test-Analyze-And-Fix (TAAF) strategies.

10.1.1 The Stage Gate Reliability Growth Plan

As described in Chapter 1, Reliability Growth can occur in all of the stage gates. The stage gate Reliability Growth plan is shown in Figure 10.2. As

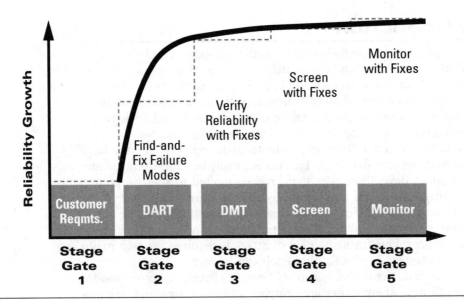

Figure 10.2
Accelerated Reliability Growth program

shown, most of the growth should occur in the first accelerated testing stage gate 2. This shows how product reliability growth is viewed and planned for over the stage gate phases or levels. The advantage of using the stage gate growth plan is that it is practical for industry.

Generally, there are two basic kinds of ARG test methods used: constant stress testing and step-stress testing. Constant stress testing applies to an elevated stress maintained at a particular level over time (see Chapters 4, 5, and 9), such as isothermal aging, in which parts are subjected to the same temperature for the entire test (similar to a burn-in). Step-stress testing can apply to such stresses as temperature, shock, vibration, and Highly Accelerated Life Test (HALT) as described in Chapter 4. These tests stimulate potential failure modes, and Reliability Growth occurs when failure modes are fixed.

No matter what the method, ARG planning is essential to avoid wasting time and money when accelerated testing is attempted without an organized program plan like a stage gate process.

10.1.2 What Is the Best Accelerated Testing Program?

There are numerous types of accelerated tests including HALT, Step-Stress, Highly Accelerated Stress Screening (HASS), Environmental Stress Screening (ESS), failure-free accelerated testing, Reliability Growth, accelerated reliability growth, and so forth (see Figure 10.3). These practices are all important, since each has been used today in both commercial and industrial applications for ensuring product reliability. The methods have not been without confusion. Confusion exists as to when and which test method should be used and the Reliability Growth that can be achieved with any one method.

In this chapter, and throughout this book, our approach subscribes to integrating and implementing these test techniques throughout the product development cycle using accelerated Reliability Growth, which is linked to stage gate levels/phases. Stage gate simply provides a format for test integration in a timely manner. Table 10.1 summarizes the approach and how these tests fit into the stage gate process.

Figure 10.3
What is the best accelerated testing program?

Table 10.1
*Accelerated Reliability
Growth stage gate tests
and methods*

Accelerated Test or Methods	Stage Gate(s) (Chapter)	Definitions and Uses
Reliability Growth	1–5 (Ch. 10)	As defined in Reference 1, Reliability Growth is the positive improvement in a reliability parameter over a period of time due to changes in product design or the manufacturing process. A Reliability Growth program is commonly established to help systematically plan for reliability achievement over a program's duration so that resources and reliability risks can be managed. The manner in which it is applied here is through stage gate combined with accelerated test methods.
Understanding Customer Requirements	1 (Ch. 2)	Understanding Customer Requirements is a common sense topic that is often overlooked. It can be a simple exercise or include a full approach. In a full approach, tools such as FMEA, competitive Benchmarking, and Reliability Predictive Modeling are used to assure the smartest approach in a product maturation program.
HALT (Highly Accelerated Life Test)	2 (Ch. 3)	HALT is a type of step-stress test that often combines two stresses, such as temperature and vibration. This highly accelerated stress test is used for finding failure modes as fast as possible and assessing product risks. Frequently it exceeds the equipment-specified limits.
Step-Stress Test	2 (Ch. 3, 6, 9)	Exposing small samples of product to a series of successively higher "steps" of a stress (like temperature), with a measurement of failures after each step. This test is used to find failures in a short period of time and to perform risk studies.
HAST (Highly Accelerated Stress Test)	3 (Ch. 3, 4, 9)	This test is performed in a sealed chamber, such as an autoclave, enabling higher-than-atmospheric pressures to occur. This allows a humid environment with temperatures above 100°C. As a result, shorter test times can be achieved. For example, a 1000-hour 85°C/85%RH test can be replaced by an 80-hour 130°C/85%RH HAST test at 33.5 psi.
Failure-Free Test	3 (Ch. 4, 9)	This is also termed zero failure testing. This is a statistically significant reliability test used to demonstrate that a particular reliability objective can be met at a certain level of confidence. For example, the reliability objective may be 1000 FITs (1 million hours MTTF) at the 90 percent confidence level. The most efficient statistical sample size is calculated when no failures are expected during the test period. Hence the name.
HASS (Highly Accelerated Stress Screen)	4, 5 (Ch. 5)	This is a screening test or tests used in production to weed out infant mortality failures. This is an aggressive test since it implements stresses that are higher than common ESS screens. When aggressive levels are used, the screening should be established in HALT testing.
ESS (Environmental Stress Screening)	4, 5 (Ch. 5)	This is an environmental screening test or tests used in production to weed out latent and infant mortality failures.

Actually the concept of mathematically linking maturation levels to a Reliability Growth curve (see Figure 10.2) has previously been described (see Reference 20). An instructive example is given in this chapter's Appendix.

10.2 Estimating Benefits with Reliability Growth Fixes

Accelerated test practices promote aggressive Reliability Growth to achieve improved reliability beyond the warranty period. This is part of the Development Phase philosophy. If failure mode fixes are incorporated into the product, Reliability Growth will be achieved. For example, typical component wear-out is greater than 25 years, while customer life specifications are often between 10 to 20 years. If a product goes through a stage gate process and incorporates fixes of observed failure modes, even conservative estimates in failure rate reduction are significant. However, if fixes are not incorporated, even with a production screening program, only the infant mortality failures can be removed and no actual improvements can be made in the steady-state failure rate which remains constant throughout customer usage. Therefore, Reliability Growth is important since a company's reputation is at stake, and although a company supplies hundreds of reliable components, "one bad apple can spoil the barrel." Thus, if you're a manager trying to cut costs, you don't want to cut your Reliability Growth program and risk going out of business!

When corrective action fixes are incorporated into products, reliability will improve significantly, lowering the risk of excessive sparing and field returns. As an example, consider a 10,000 FIT (100,000 hours MTBF) assembly. Let us assume that 95 percent of the potential failures can be assigned corrective action fixes. Historically, fix effectiveness factors range from 60 percent to 80 percent with an average of 70 percent. On average, we anticipate being able to improve the product failure rate by 66.5 percent (= 0.7 of 95 percent). As a result, products designed for 10,000 FITs can exceed the reliability objective with corrective action fixes. Using these estimates, the failure rate will be reduced down from 10,000 FITs to ~3,350 (300,000 hours MTBF). This is a factor of 3 improvement (that is $1/(1 - 0.665)$ divided by 10,000 FITs). (To clarify, a failure rate of 10,000 FITs is 10 failures per one million hours. Assuming a 66.5 percent improvement means that 66.5 percent of the failure modes can be removed. Effectively this is 0.665 times 10 failures per million hours or 6.65 failures removed per million hours leaving 3.35 failures per million hours. And 3.35 failures per million hours is 3,350 FITs.)

Figure 10.4
Accelerated Reliability Growth saves money

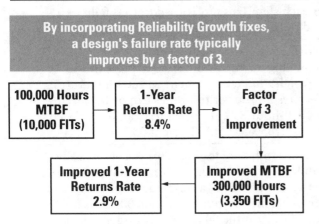

Disregarding the time it takes to incorporate the fixes, reliability, *R*, can improve each year from (see Chapter 8)

$$R \ (10{,}000 \text{ FITs}) = 0.916 \quad \text{to} \qquad\qquad (10.1)$$
$$R \ (3{,}350 \text{ FITs}) = 97\%$$

This is shown in Figure 10.4. The growth factor is 3 (=10,000/3,350). It implies that field failure and related problems such as excessive sparing should improve from 8.4 percent to 3 percent per year for the example cited. Using these estimates, Table 10.2 provides an overview of estimated reliability benefits of incorporating a Reliability Growth program.

Reliability Objective FITs (MTBF)	Reliability (Per Year)	Sparing (Per Year)	Improved* Growth Failure Rate FITs (MTBF)	Improved* Growth Reliability (Per Year)	Reduced* Sparing Estimate (Per Year)
1000 (1×10^6 Hrs.)	0.991	0.9%	335 (3×10^6 Hrs.)	0.997	0.3%
4000 (2.5×10^5 Hrs.)	0.966	3.4%	1,340 FITs (7.5×10^5 Hrs.)	0.988	1.1%
10,000 (1×10^5 Hrs.)	0.916	8.4%	3,350 FITs (3×10^5 Hrs.)	0.971	2.9%
40,000 (2.5×10^4 Hrs.)	0.704	29.6%	13,400 FITs (7.5×10^4 Hrs.)	0.889	11.1%

*Estimates are based on a 66.5% improvement or 33.5% reduction in failure rate.

Remember that a Reliability Growth program should be used even in screening and monitoring. Screening and monitoring without incorporating fixes does not grow reliability. Figure 10.5 illustrates the concept. Screening and monitoring are discussed in Chapter 5.

Table 10.2
Estimated conservative benefits of a Reliability Growth program

10.3 Accelerated Reliability Growth Methodology

The basic methodology in ARG planning should minimally consist of the following (see References 1–3):

- The design of appropriate accelerated tests to stress expected or unexpected failure modes of the subsystems. These tests should be chosen and designed to stimulate failures at a faster rate. An effective program will include a streamlined root-cause corrective action plan. Without a complete plan, accelerated testing will be wasted and Reliability Growth opportunities lost.

- The correct use of historical acceleration factors and Reliability Growth parameters (e.g., alpha). This will enable estimates of accelerated Reliability Growth over the program's testing phases to be generated.

- The accurate tracking and assessment of Reliability Growth during and after each test phase and corrective action. This aids in correct assessment techniques for fix-effectiveness to properly evaluate compliance of growth goals/milestones. Planning Reliability Growth is not enough; periodic assessments of reliability are also essential so that management is assured that their achievement of the planned Reliability Growth goals are realistic.

Figure 10.5
Production Reliability Growth screening and monitoring program

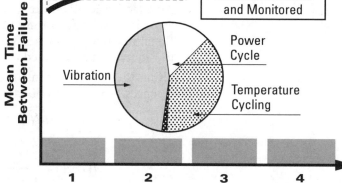

It is important, when possible, to use the idealized accelerated Reliability Growth equations and curves; these are integral to Reliability Growth and in establishing interim goals. They provide target accelerated test times and aid in the estimation of the expected number of failures for each phase of the sub-system's qualification program.

After testing, an accelerated test failure should be subjected to the standard failure analysis procedures. Failed parts should be part of a Failure Review and Corrective Action System for post-test examination to identify failure mechanisms. The ARG plan should include guidelines to properly determine each root cause and corrective action. A Corrective Action Review Team should be organized to efficiently review solutions to each problem. After corrective actions have been implemented, Reliability Growth can be assessed. Assessment, Failure Review and Corrective Action System, Corrective Action Review Team, and other processes are important Reliability Growth tools for management to ensure that goals are met.

If tests are not designed correctly to initiate all or most of the product's potential failure modes, major problems will result, causing delays in the program's progress. Programs often neglect such details, e.g., seeing that corrective actions are properly assigned to each fix in root-cause failure analysis.

10.4 Applying Accelerated Reliability Growth Theory

Integral to ARG planning are idealized Accelerated Reliability Growth equations and curves that characterize realistic growth for complex subsystems across the major test phases. They are useful in representing total program growth, determining whether test duration is sufficient to achieve the reliability requirements, estimating average Mean Time Between Failure (MTBF) over each phase, and estimating growth rates. They offer important flexibility in the planning stages. For example, growth rates are unknown in the initial planning stages. Using conservative historical growth "alphas" and estimating initial MTBF subsystem values, however, can still project Reliability Growth. If accelerated testing is employed at the subsystem level, Reliability Growth can still be estimated over the testing phases by conservatively estimating time compression. Just as conservative estimates of growth alphas are known, available historical data often enable one to estimate an effective acceleration factor for test planning purposes. Under the following assumptions (see References 2 and 3), ARG planning can be performed:

- An effective acceleration factor, A, exists and can be estimated.
- Time is linearly compressed by this factor A.
- Equal reliability growth is possible in an uncompressed time period, as in the equivalent accelerated compressed time period.

Using these assumptions, the idealized Reliability Growth equations (see References 1–3) are described in Equation 10.2.

$$M(t) = M_I \qquad\qquad t \le t_1$$

$$M(t, A) = \frac{M_I}{1 - \alpha} \left(\frac{t}{t_1}\right)^{\alpha} A^{\alpha} \qquad t \ge t_1$$

(10.2a)

where

$A\ $ = the effective acceleration factor,

M_I = the initial MTBF,

$\alpha\ $ = the growth parameter,

$t\ $ = the cumulative test time under accelerated conditions, and

t_1 = the cumulative test time over the first phase.

These equations are used when acceleration is applied to all phases including Phase 1. The equation becomes slightly modified if accelerated testing starts after t_1 (see References 2 and 3) and is

$$M(t, A) = \frac{M_I}{1 - \alpha} \left(\frac{t}{t_1}\right)^{\alpha} A^{\alpha} \left\{1 - \frac{t_1}{t_f}\right\} \qquad t_1 < t < t_f \qquad (10.2b)$$

However, this additional factor turns out to be close to unity for most applications since $t_f \gg t_1$. Compared to the idealized Reliability Growth equations (see Reference 1), Equation 10.2 is modified simply by the effective acceleration factor. In the event that A is equal to one, the standard idealized reliability growth equations result (see References 1–3).

▼ **Example 10.1** *Predicting Accelerated Reliability Growth*

Problem:

Consider a new subsystem with an estimated initial MTBF of 14,000 hours. A customer wants to buy this subsystem from your company but is requesting an MTBF of 200,000 hours. Estimate if this much Reliability Growth is possible. Use a conservative growth alpha of 0.25. Assume that no more than 3,000 hours of accelerated testing can be used to find and fix failure modes for growing this product's reliability. Use 200 as a best estimate for the maximum reasonable testing acceleration factor.

Solution:

To provide an estimate, first estimate t_1 using the initial MTBF value [2, 3] as

$$t_1 = \frac{3 \, M_I}{A} = \frac{3 \, (14,000)}{200} = 210 \qquad (10.3)$$

Here, we use a factor of 3 times M_I. This factor ensures a 95 percent chance of observing a failure in the initial test period (t_1) where

$$1 - R(t_1) = 1 - Exp\{-3 \, t_1/MTBF\} = 1 - Exp\{-3 \, M_I/M_I\} = 95\%$$

Next, insert these values into the ARG equation using an historic conservative growth alpha of 0.25 to obtain

$$MF = \frac{14,000}{1 - 0.25} \left(\frac{3000}{210}\right)^{0.25} 200^{0.25} = 136,473 \quad hours \qquad (10.4)$$

The results indicate that it will be initially difficult to meet the customer's MTBF reliability requirement of 200,000 hours and that some risk mitigation will be required to proceed (see Evaluating Product Risks, Chapter 13). This result can be presented graphically to the customer. For example, tools such as Microsoft® Excel can be used to program a spreadsheet and graphically display growth curves. Table 10.3 shows the Excel input/output table from this program, and Figure 10.6 displays the graphical output (also see References 2 and 3).

Figure 10.6 illustrates the idealized curves for a subsystem using the ARG equations above. The MTBF at each point of a Test-Analyze-And-Fix Reliability Growth phase is shown. After each phase, corrective actions are incorporated into the subsystem, yielding a jump in MTBF. The acceleration factor (*AF*) is estimated from accelerated conditions and expected typical failure modes. A conservative historic growth alpha of 0.25 should be used unless applicable data are available.

Inputs Required	Symbol	Selected Input	Output Plot Times	Output MF Final Unaccel.	MFA Final with Accel.	Average MFA with Accel.	Phase	MFA/M$_I$
MTBF Start Value	M$_I$	14,000	0	14,000	14,000	14,000	1	1
			210	14,000	14,000	14,000	1	1.0
Phase 1 Start Value	t$_1$	210	210	18,667	70,198	79,308	2	5.0
			307	20,520	77,168	79,308	2	5.5
Phase 2 Ends	t$_2$	500	403	21,972	82,629	79,308	2	5.9
			500	23,188	87,199	79,308	2	6.2
Phase 3 Ends	t$_3$	1,000	500	23,188	87,199	95,923	3	6.2
			667	24,917	93,702	95,923	3	6.7
Phase 4 Ends	t$_4$	2,000	833	26,343	99,067	95,923	3	7.1
Phase 5 Ends	t$_5$	3,000	1,000	27,575	103,698	114,072	4	7.4
			1,333	29,631	111,431	114,072	4	8.0
Reliability Growth Alpha	Alpha	0.25	1,666	31,328	117,812	114,072	4	8.4
			2,000	32,792	123,318	114,072	4	8.8
Estimated System Level Acceleration Factor	AF	200	2,000	32,792	123,318	130,118	5	8.8
			2,333	34,081	128,163	130,118	5	9.2
			2,666	35,235	132,506	130,118	5	9.5
			3,000	36,290	136,474	130,118	5	9.7

Table 10.3
Excel spreadsheet example

Figure 10.6
Accelerated Reliability Growth prediction

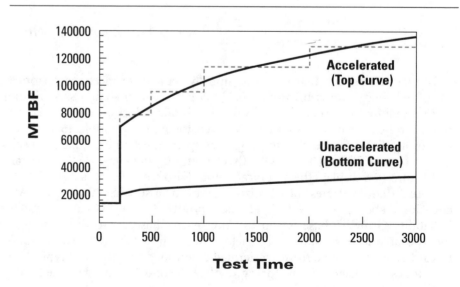

10.5 Assessing Reliability Growth

Evaluating Reliability Growth can be done in two ways. The first method utilizes assessments (quantitative evaluations of the current reliability status) that are based on information from the detection of failure sources. Intermittent program data are invaluable and very practical in ARG planning. The second method is qualitative assessment performed by monitoring various activities in the process to ensure that activities are being accomplished in a timely manner and that the effort and quality comply with the program plan (see Chapter 13). The methods complement each other in controlling the growth process. It is always best to make a quantitative evaluation if possible.

10.5.1 Method of Assessment

Assessing the effectiveness of engineering fixes for reliability would clearly enhance the ability to plan and manage a Reliability Growth program. Validation of product progress is a key issue in meeting reliability objectives. In practice, there are generally not enough failure data both before and after a corrective action to estimate with reasonable confidence the effectiveness of the fix and the product's improved failure rate. This will most likely be true when implementing an accelerated testing program. Consequently, this lack of information may result in the assignment of unrealistic fix-effectiveness factors, overly optimistic or pessimistic failure rate estimates, and a corresponding incorrect assessment of reliability achievements.

There are a number of statistically sound practices in industry. For example, failure-free test methods can be employed (see Chapter 5). Alternatively, methodologies for assessing Reliability Growth from corrective actions are shown in References 17 and 18. These methodologies require modification to incorporate the effects of time compression when using accelerated testing (see References 2 and 3). Once properly modified, they can be implemented during any Reliability Growth test phase. It is often necessary to estimate a "fix effectiveness factor" that is defined as the percent decrease in a problem failure mode due to a corrective action. This is used in estimating a product's improved failure rate. These are some of the alternative solutions available when it is prohibitive to perform statistical validation of a failure rate due to large sample size and test time requirements.

10.6 Summary

Accelerated Reliability Growth (ARG) concepts are not difficult. The important points to keep in mind are:
- ARG is designed to find failure modes as fast as possible.
- Growth occurs when we incorporate fixes.
- Statistically significant tests are designed with science in mind to verify Reliability Growth.
- Sophisticated tools are available to look at both common and uncommon problems.
- Product Reliability Growth is a responsibility that we all share to provide customer satisfaction.
- The Accelerated Reliability Growth stage gate process provides a management plan.

References

1. *Military Handbook-189*, Reliability Growth Management, 13 February 1981.

2. Feinberg, A. A., "Accelerated Reliability Growth Models," *Journal of the Institute of Environmental Sciences*, 1994, pp. 17-23.

3. Feinberg, A. A., and Gibson, G. J., "Accelerated Reliability Growth Methodologies and Models," *Recent Advances in Life-Testing and Reliability*, edited by N. Balakrishnan, CRC Press, Boca Raton, FL, 1995.

4. Peck, D. S., and Trap, O. D., (1978), *Accelerated Testing Handbook*, Technology Associates, Revised 1987.

5. Peck, D. S., and Zierst, C. H., Jr., "The Reliability of Semiconductor Devices in the Bell System," *Proceedings of the IEEE*, Vol. 62, No. 2, February 1974, pp. 260-273.

6. Reynolds, F. H., "Accelerated Test Procedures for Semiconductor Components," *15th Annual Proceedings on Reliability Physics*, 1977, pp. 168-178.

7. IEEE index (1988), *The 1988 Index to IEEE Publications*, IEEE Service Center, P.O. Box 1331, Piccatanny, NJ 08855, (201) 981-1396 and 9535.

8. Howes, M. J., and Morgan, D. V., Eds., (1981), "Reliability and Degradation of Semiconductor Devices and Circuits," *The Wiley Series in Solid State Devices and Circuits*," Vol. 6, Wiley, New York.

9. Nelson, W., *Accelerated Testing*, Wiley, New York, 1990.

10. Gibson, G. J., and Crow, L.H., "Reliability Fix Effectiveness Factor Estimation," *Annual Reliability and Maintainability Symposium*, 1989, p. 75 (89 RM-075).

11. Moura, E. C., "A Method to Estimate the Acceleration Factor for Subassemblies," *IEEE Transactions on Reliability*, Vol. 41, No. 3, 1992, p. 396.

12. Seager, J. D., and Fieselman, C. D., "A Method to Predict an Average Activation Energy for Subassemblies," *IEEE Trans. Reliability*, Vol. 37, December 1988, pp. 458-461.

13. Schinner, J. D., "Reliability Growth Through Application of Accelerated Reliability Techniques and Continued Improvement Processes," *Proceedings of the Institute of Environmental Sciences*, 1991, pp. 347-354.

14. Duane, J. J., "Learning Curve Approach to Reliability Modeling," *IEEE Transactions on Aerospace*, 2, 1964, p. 563.

15. Seusy, C. J., "Achieving Phenomenal Reliability Growth, *ASM Conference on Reliability*," Key to Industrial Success, Los Angeles, CA, pp. 24-26 March, 1987.

16. Hobbs, G., "Highly Accelerated Life Testing," *Proceedings of the Institute of Environment Sciences*, 1992, pp. 377-381.

17. Gibson, G. J., and Crow, L. H., "Reliability Fix Effectiveness Factor Estimation," *Annual Reliability and Maintainability Symposium*, 1989, p. 75 (89 RM-075).

18. Crow, L. H., "Methods of Assessing Reliability Growth Potential," 1984, pp. 48-189.

19. Feinberg, A. A., Gibson, G. J., and Shupe, R. H., "Connecting Technology Performance Maturation Levels to Reliability Growth," *Proceedings of the Institute of Environmental Sciences*, 1992, pp. 415-421.

APPENDIX

Accelerated Reliability Growth Stage Gate Model

It is possible to link the Reliability Growth curve directly to stage gate levels (or phases). Mathematically linking maturation levels to a Reliability Growth curve in this way has previously been described (see Reference 20). In this Appendix, we provide an example of this procedure for the interested reader. This exercise may appear somewhat academic, as traditional Reliability Growth is difficult to apply over phases with unknown acceleration factors and different tests. However, even in traditional Reliability Growth, it is often customary to assign growth goals to developmental stages in a program (see Reference 1). Thus, in reality, a qualitative growth model can be planned and mapped in a traditional sense to development levels. One advantage of attempting to estimate growth is in assessing a project's risk. If conservative growth estimates indicate that achieving a reliability goal would be unlikely, then a high risk can be assigned which will most likely affect the management of the project. As an example, we will take a traditional Reliability Growth example cited in Military Handbook-189 (see Reference 1, page 46). This example has also been described in the Accelerated Reliability Growth case in References 2, 3. The results are shown in Figure 10.A.1 (see Reference 2). The model for the accelerated growth curve is given in Equation 10.2. The growth parameters for this example are $\alpha = 0.23$, $M_1 = 50$, $t_1 = 1000$, with an overall acceleration factor of $A = 20.36$. Note, there are five test phases which make it a good example for mapping the five-phase stage gate process in this book. Also, in Figure 10.A.1, the MTBF has been put in units of 1,000 hours for a more typical expected commercial case. The test time phases are $t_2 = 2500$, $t_3 = 5000$, $t_4 = 7000$, and $t_5 = 10000$. These test phases are not unreasonable for a stage gate process. For example, Phase 3 is actually 2,500 hours in length (= 5,000 – 2,500 hours). This could easily be the length of a design maturity test. Here we might wish to refine an estimate of the growth potential over a particular stage gate in which growth factors can be more accurately assessed.

Figure 10.A.1
Accelerated and unaccelerated Reliability Growth curve (see References 2 and 3)

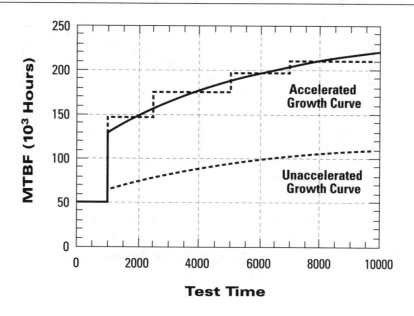

To map this planned growth to a five-level stage gate process, we use the simple mapping theory described in Reference 20. In this reference, Reliability Growth is mapped to a program's development levels, L, as

$$M(t) = M_I \qquad\qquad L \le L_1$$

$$M(t, A) = \left(\frac{L_i}{L_1}\right)^{\gamma+1} M_1 \qquad L \ge L_1$$

(10.A.1)

To find gamma (as shown in Reference 20), one simply equates

$$\frac{M(t)}{M_1} = \left(\frac{L_i}{L_1}\right)^{\gamma+1}$$

(10.A.2)

We wish to map the growth curve in Figure 10.A.1. The curve actually starts at a value of about 130 and increases to 221 hours (See Table 1, Reference 3). Since we are using a five-level stage gate process, with Level $L_1 = 1$ and the ith top level of 5, we have

$$\frac{221}{130} = \left(\frac{5}{1}\right)^{\gamma+1}$$

(10.A.3)

Solving, we find that $\gamma + 1 = 0.33$. Then the curve is mapped to the five-level stage gate process as

$$M(t) = 50 \qquad\qquad L_i \le 1$$

$$M(t, A) = \left(\frac{L_i}{1}\right)^{0.33} 130 \qquad L_i \ge 1$$

(10.A.4)

Figure 10.A.2
Mapped stage gate Accelerated Reliability Growth curve of Figure 10.A.1

The result is shown in Figure 10.A.2. As envisioned in our model, most of the growth is expected to occur in stage gate 2.

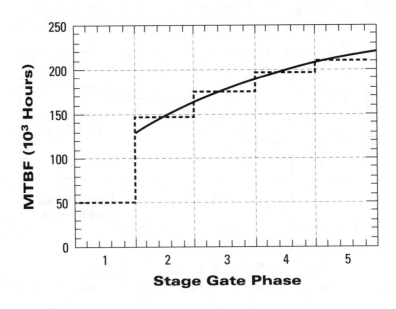

CHAPTER 11

Reliability Predictive Modeling

11.1 Introduction

Reliability predictions play a critical role in the Design for Reliability (DfR) process during product development. Reliability predictions provide early estimates of the design complexity that relate to the product reliability. When a prediction method is accompanied by appropriate realism factors, it can also provide excellent estimates of the expected reliability in actual use conditions.

Reliability predictions are generally made for steady-state operation. The steady-state portion of life is discussed in Chapter 8, and it is depicted in this chapter by the bathtub curve in Figure 8.6. The steady-state region is modeled with a constant failure rate. Reliability predictions can be performed for any aspect of the bathtub curve or for any other realistic characteristics, but this chapter is focused on reliability predictions for this constant failure rate region.

Reliability predictions can be used for many purposes during product development. Typical applications include:
- determining the feasibility of meeting a reliability requirement or a goal;
- monitoring the complexity during the development process;
- estimating the expected rate of failures for the associated life-cycle cost;
- estimating the failure rates that support design trade-off evaluations;
- estimating failure rates for calculating failure rate-dependent characteristics such as maintainability or testability;
- estimating the failure rates for a Failure Modes and Effect Analysis (FMEA);
- supporting a customer-requested evaluation; and
- providing the failure rate expectations for various conditions (e.g., thermal extremes or user environments such as mobile and fixed ground sites).

11.2 System Reliability Modeling

Most common reliability predictions use a "bottoms-up" approach by estimating the failure rate for each element and then combining the failure rates for the entire assembly (see Figure 11.1). In the block diagram configuration, the system is broken down to the lowest elements of interest. Figure 11.1 illustrates a number of important block diagram representations. Here, the system failure rate is the sum of the individual subsystems A, B, C, and D. The subsystems are in a series configuration; if any subsystem fails, it results in a system failure. There are three traditional types of block diagrams represented. The subsystems shown in Figure 11.1 have elements that are purposely configured to illustrate each type. Subsystem A consists of parts 1, 2, and 3 that

Figure 11.1
System reliability block diagram representation

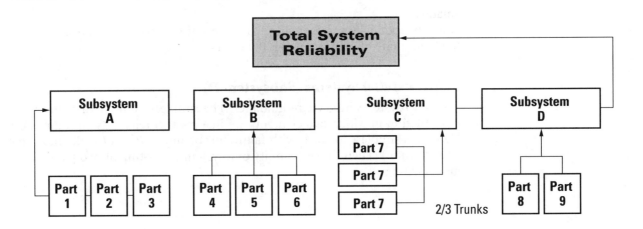

are in the series configuration (if any part fails, then subsystem A fails). Subsystem B consists of parts 4, 5, and 6 in a parallel configuration. Similarly, subsystem D has parts 8 and 9 in parallel. Parallel subsystems indicate redundancy. For example, in subsystem B, any two parts can fail without failure to the subsystem. All three would have to fail for subsystem B to fail. Lastly, subsystem C has elements that appear to be in parallel. However, the 2/3 trunk is used to indicate that as long as two of the three elements are working, the subsystem is operational.

Once the failure rate for each subsystem is determined, the results can be rolled up into a reliability prediction for the system itself. This is the bottoms-up approach. In each type of configuration, the method for determining the failure rate is different.

11.2.1 Series Systems (Subsystem A)

Reliability predictions are generally made for steady-state operation. Therefore, usually the exponential distribution is assumed (see Figure 8.7). The reliability function for the exponential distribution

$$R(t) = e^{-\lambda t} \qquad (11.1)$$

is the probability of a component surviving to time, t. Therefore, for a system made up of n independent components, where any component failure causes a system failure, the probability of survival for the whole system is

$$R_{system} = R_1 R_2 R_3 R_4 \cdots R_n \qquad (11.2)$$

or

$$e^{-\lambda_{system} t} = e^{-\lambda_1 t} e^{-\lambda_2 t} e^{-\lambda_3 t} \cdots e^{-\lambda_n t} = e^{-(\lambda_1 + \lambda_2 + \lambda_3 + \cdots + \lambda_n)\, t} \qquad (11.3)$$

Therefore,

$$\lambda_{system} = \lambda_1 + \lambda_2 + \lambda_3 + \cdots + \lambda_n$$

The failure rate of the system is simply the sum of the failure rates of the individual devices.

▼ **Example 11.1** *Failure rate of subsystem A*

Problem:
The failure rates of parts 1, 2, and 3 are 0.1, 0.3, and 0.5 hours^{-1}, respectively. Determine the failure rate for subsystem A.

Solution:
Since the total failure rate is simply the sum of the individual rates, the failure rate of subsystem A is 0.9 hours^{-1}.

11.2.2 Parallel Systems (Subsystem D)

Parallel systems indicate redundancy. The simplest example of redundancy is a situation in which two elements are in a parallel reliability configuration. Here, it is simplest to work with failure probabilities $R' = (1 - R)$ (indicated with a prime). Then the probability of a system consisting of two parallel elements failing is

$$R'_{system} = R'_1 R'_2 \qquad (11.4)$$

or
$$1 - R_{system} = (1 - R_1)(1 - R_2) \qquad (11.5)$$
Solving this gives
$$R_{system} = R_1 + R_2 - R_1 R_2 \qquad (11.6)$$
Substituting in the reliability function for the exponential distribution yields

$$R_{system} = e^{-\lambda_1 t} + e^{-\lambda_2 t} - e^{-\lambda_1 t} e^{-\lambda_2 t} = e^{-\lambda_1 t} + e^{-\lambda_2 t} - e^{-(\lambda_1 + \lambda_2)t} \qquad (11.7)$$

The right-hand side is a complicated function, and R_{system} cannot be put into simple exponential form. Therefore, the actual system failure rate cannot easily be defined by the exponential distribution. However, both sides can be integrated over all time as
$$(11.8)$$

$$\int_0^\infty R_{system}(t)dt = \int_0^\infty [e^{-\lambda_1 t} + e^{-\lambda_2 t} - e^{-(\lambda_1 + \lambda_2)t}]dt = \int_0^\infty e^{-\lambda_1 t}dt + \int_0^\infty e^{-\lambda_2 t}dt - \int_0^\infty e^{-(\lambda_1 + \lambda_2)t}dt$$

The left-hand side is the system's MTTF (see Section 8.2)

$$MTTF_{system} = \int_0^\infty R(t)\ dt = \frac{1}{\lambda_{eff}} \qquad (11.9)$$

and its inverse, λ_{eff}, is denoted as an effective failure rate for the system. The integrals on the right-hand side of the equation have constant failure rates. For example,

$$\int_0^\infty e^{-\lambda_1 t}\ dt = \frac{1}{\lambda_1} \qquad (11.10)$$

Then a system's MTTF for two elements in parallel is

$$MTTF_{system} = \frac{1}{\lambda_{eff}} = \frac{1}{\lambda_1} + \frac{1}{\lambda_2} - \frac{1}{\lambda_1 + \lambda_2} \qquad (11.11)$$

▼ **Example 11.2** *Failure rate of subsystem D*

Problem:
The failure rates of parts 8 and 9 in subsystem D are 0.25 and 0.2 hour^{-1}, respectively. Determine the effective failure rate of subsystem D.
Solution:
The failure rate for subsystem D is found from Equation 11.11

$$\frac{1}{\lambda_{eff\ D}} = \frac{1}{\lambda_8} + \frac{1}{\lambda_9} - \frac{1}{\lambda_8 + \lambda_9}$$

Substituting in the failure rate values gives

$$MTTF_D = \frac{1}{\lambda_{eff\ D}} = 4 + 5 - 2.22 = 6.78$$

Therefore, the effective failure rate of subsystem D is 0.147 hour^{-1}.

Figure 11.2

Reducing redundancy in modeling

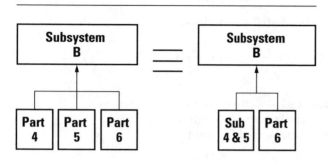

11.2.3 Reducing Redundancy in Modeling

In general, the above approach can be used for a subsystem with three items in parallel, such as subsystem B in Figure 11.1. However, this can become cumbersome. Alternately, redundancy can be reduced, allowing the use of the formula above. For example, modeling redundancy of subsystem B can be reduced as shown in Figure 11.2.

Here, combining parts 4 and 5 creates a new subsystem. The effective failure rate of this subsystem is that of two parts in parallel given by

$$\frac{1}{\lambda_{eff\ Sub\ 4\&5}} = \frac{1}{\lambda_4} + \frac{1}{\lambda_5} - \frac{1}{\lambda_4 + \lambda_5} \qquad (11.12)$$

This reduces the problem of determining the failure rate for the overall subsystem. The failure rate for subsystem B is now reduced to a problem of determining the failure rate for two elements in parallel and is given by

$$MTTF_D = \frac{1}{\lambda_{eff\ B}} = \frac{1}{\lambda_{eff\ Sub\ 4\&5}} + \frac{1}{\lambda_6} - \frac{1}{\lambda_{eff\ Sub\ 4\&5} + \lambda_6} \qquad (11.13)$$

▼ **Example 11.3** *Failure rate of subsystem B*

Problem:
The failure rates of parts 4, 5, and 6 in subsystem B are 0.2, 0.4, and 0.25 hour^{-1}, respectively. Determine the effective failure rate for subsystem B.

Solution:
Combining parts 4 and 5 into a subsystem yields (Equation 11.12)

$$\frac{1}{\lambda_{eff\ Sub\ 4\&5}} = 5 + 2.5 - 1.66 = 5.83$$

This gives an effective failure rate of 0.171 hour^{-1}. Then the effective failure rate of subsystem B is found from Equation 11.13 as

$$\frac{1}{\lambda_{eff\ Subsystem\ B}} = 5.83 + 4 - 2.375 = 7.46$$

This gives the effective failure rate of subsystem B of 0.134 hour^{-1}.

11.2.4 Modeling k of n Subsystem Elements

The probability of at least k out of n identical elements working in a subsystem is given in probability theory. In terms of the probability of success R, this is

$$R_{system} = \sum_{i=0}^{n-k} \frac{n!}{i!(n-i)!} (R_A)^{n-i} (1 - R_A)^i \qquad (11.14)$$

▼ **Example 11.4** *Failure Rate of subsystem C and the system*

Problem:
Determine the effective failure rate for subsystem C in which at least 2 out of 3 items must be working. The failure rate of item 7 is constant and is 0.2 hours^{-1}. Then using this result and that of Examples 11.1 through 11.3, determine the effective failure rate of the system.

Solution:
Using Equation 11.14 yields

$$R_{Subsystem\ C} = \frac{3!}{0!(3-0)!}(R_7)^3(1) + \frac{3!}{1!(3-1)!}(R_7)^2(1-R_7) = (R_7)^3 + 3(R_7)^2(1-R_7)$$

In terms of the exponential reliability function, this is

$$R_{Subsystem\ C} = e^{-3\lambda_7 t} + 3e^{-2\lambda_7 t}(1 - e^{-\lambda_7 t})$$

Simplifying this is

$$R_{Subsystem\ C} = 3e^{-2\lambda_7 t} - 2e^{-3\lambda_7 t}$$

Integrating both sides of this equation over all time gives

$$MTTF_{Subsystem C} = \frac{1}{\lambda_{eff\ C}} = \frac{3}{2\lambda_7} - \frac{2}{3\lambda_7} = \frac{5}{6\lambda_7}$$

Substituting in the failure rate for part 7 of 0.2 hours^{-1} gives

$$\lambda_{eff} = \frac{6\lambda_7}{5} = \frac{6(0.2)}{5} = 0.24$$

From Examples 11.1 through 11.4, the effective failure rates for subsystems A, B, C, and D have now been determined as 0.9, 0.134, 0.24, and 0.147 hours^{-1}, respectively. Therefore, the total effective failure rate of the system can be obtained from the sum as 1.42 hours^{-1} (or a 0.704-hour system MTTF).

11.2.5 Other Configurations and Repair/Availability

This section has covered a number of typical block diagram configurations. Other configurations exist. Appendix A covers a number of k out of n type redundant configurations. Additionally, the effective failure rate for a redundant system can be extended with repair. That is, when one item fails in a redundant system, there is an opportunity to repair it before subsystem failure. Therefore, the effective failure rate needs to incorporate the possibility that the unit will be fixed and back online with full redundancy again available before subsystem failure. Appendix B tabulates these types of situations. Additionally, sometimes redundancy is achieved by having units on standby waiting to be substituted for a potential component failure. In this case, the effective failure rate is a function of the switching mechanism and the number of active units. This is also described in Appendix B.

For very complex systems that require frequent repair, often the reliability metric of interest by a customer, is expected equipment availability. This is described in Appendix C.

11.3 Customer Expectations

Customer expectations for reliability predictions can vary quite significantly, especially with a global market that encompasses a wide range of applications. Therefore, it is important to be prepared to respond to a remarkably wide range of expectations. It is common for multiple methods to be performed during a development project. In fact, it would be unusual if only one reliability prediction was performed per project.

11.4 Various Methods

There are many available methods for performing reliability predictions in our industry. Each method has some advantages and some weaknesses. This section will describe some of the frequently applied methods in sufficient detail, to understand what is involved in performing each method and what type of output is available. Many methods have been omitted to focus on predominant methods.

There are many excellent tools available. Expertise with a specific tool will substantially impact the user's satisfaction and productivity, regardless of the tool or the prediction method. This chapter focuses on two specific methods:
- Military Handbook-217 (latest versions FN2) for both Parts Count and Detail Stress Methods to estimate electrical and electronic parts, using an exponential failure rate, and
- Bellcore (latest issue 6) for Methods I, II, and III assuming a serial model, using exponential failure rates of electrical parts.

An example that compares Military Handbook-217 to Bellcore is provided in this chapter. Some discussions are provided on related materials and techniques for:
- estimating system reliability where the system elements are configured in six typical redundancy relationships, using the relationships defined in Reference 1;
- estimating extremely complex redundancy using discrete event simulation techniques;
- estimating acceleration factors using the Arrhenius relationship;
- estimating acceleration factors using a wide range of other application factors; and
- estimating the resulting probability based on a combination of a wide range of conditional probabilities.

11.4.1 Military Handbook-217 Predictive Methods

In this section, Military Handbook-217 is discussed, versions E, F-1, F-2, Parts Count and Detail Stress Methods, to estimate failure rates of electrical and electronic parts using an exponential distribution. Military Handbook-217 has the most internationally recognized methods. For example, the Russian standard for reliability predictions can be read, even for one who cannot read Russian.

Versions of Military Handbook-217 have been widely used for more than 30 years. The major advantage of the Military Handbook is the widespread application and experience many people have with some type of realism factors, from a comparison of past predictions to the actual experience under some specific use conditions. The latest revision is Notice 2 for Military Handbook-217F, *Military Handbook Reliability Prediction of Electronic Equipment*. The Military Handbook describes both the Parts Count and Detail Stress Methods. The Military Handbook also provides failure rates for many environmental conditions, covering the range from ground benign for continuous operation in comfort-controlled conditions to cannon launch for electronics.

The Parts Count Method makes assumptions for a representative thermal ambient, part complexity, and various electrical stresses. These assumptions simplify the effort to perform these evaluations. This simplification allows for early evaluations to be performed. The Parts Count Method is ideal if the design is at a very early stage or if the analysis labor is to be minimized. Section 11.4.3 provides an example of the Military Handbook-217 Parts Count Method for a commercial electric clock. A Detail Stress Method uses the specific parts complexity and the specific application stresses. This added detail requires more time for the collection of parts library information and application stresses. The advantage of the Detail Stress Method is that the output conclusions will reflect the specific conditions and include the effects of the thermal conditions on the failure rate. This is very helpful if the usage includes multiple conditions or if accelerated testing is planned. The primary limitation for either the Parts Count or the Detail Stress Methods is the availability of realism factors or experience comparing past reliability predictions with actual experience. Great care must also be exercised in selecting and interpreting the outputs in terms of the versions. For example, in Version F, Notice 2, Detail Stress Method has a much higher failure rate than Version F, Notice 1. Finally, note that the units used with Military Handbook failure rate is in Failures Per Million Hours (FPMH).

11.4.2 Bellcore Predictive Methods

Bellcore is the research group for AT&T. They created the Bellcore methods because they were not satisfied with the applicability of the Military Handbook methods for their commercial products or for their markets. They created the reliability prediction guidance documents for use on their products. Bellcore is intended for commercial (i.e., nonmilitary) parts. The latest revision is Technical Reference TR-332 Issue 6, December 1997, called *Reliability Prediction Procedure for Electronic Equipment*. In this section, Bellcore Methods I, II, and III are discussed, assuming a serial model for exponential failure rates of electrical parts. Table 11.1 provides an overview of the differences between Bellcore prediction methods and Military Handbook-217.

Their document says one of the purposes is for the recommended failure rates to contain the appropriate realism experience. Most practitioners who have done extensive comparisons will advise you to confirm the realism on your own products in their application. In fact, Methods II and III address this issue.

Method I uses a very similar approach to the Military Handbook-217 Parts Count Method. It uses representative complexities, stresses, and one or more environments as the basis. It only considers three environments: controlled fixed ground, uncontrolled fixed ground, and mobile ground. It addresses four quality levels: 0, I, II, or III. Additional factors translate from the representative conditions to other specific conditions, if desired. A Bellcore Method I prediction for a commercial electric clock is provided in the next section. Method II is based on combining Method I predictions with data from a laboratory test performed in accordance with specific Bellcore test criteria. Method III is for statistical predictions of in-service reliability based on field tracking data collected in accordance with specific Bellcore criteria. The Bellcore failure rate output is in units of FITs, which is equivalent to failures per billion hours (see Chapter 8).

The primary limitation for Method I is the availability of realism factors or experience comparing past reliability predictions with actual experience. Great care must be exercised in selecting and interpreting the outputs in terms of the versions.

Prediction Type	Conditions		
Typical Procedure	Information	Environmental Factors	Quality Factors
Military Handbook-217 (Parts Count)	Assume 40°C op. temp. and 50% electrical stress	Ground benign, fixed, or mobile. Airborne, inhabited cargo, missile launch, etc.	Jantxv, Jantx, Jan Commercial, Plastic
Military Handbook-217 (Detailed Stress)	Specify stress level each component	Ground benign, fixed, or mobile. Airborne, inhabited cargo, missile launch, etc.	Jantxv, Jantx, Jan Commercial, Plastic
Bellcore Method I, case 1 (Parts Count)	No device burn-in or unit burn-in <1 hr, 40°C, 50% stress	Ground fixed or mobile, airborne, space-based commercial.	Level 0, Level I Level II, Level III
Bellcore Method I, case 2 (Parts Count)	No device burn-in and unit burn-in > 1 hr, 40°C, 50% stress	Ground fixed or mobile, airborne, space-based commercial.	Level 0, Level I Level II, Level III
Bellcore Method I, case 3 (Parts Count)	General case, anything other than 40°C, 50% stress	Ground fixed or mobile, airborne, space-based commercial.	Level 0, Level I Level II, Level III
Bellcore Method II (Combined lab data & Parts Count)	Predictions based on combined Parts Count and lab data	Ground fixed or mobile, airborne, space-based commercial.	Level 0, Level I Level II, Level III
Bellcore Method III (Predictions from field tracking)	Prediction based on field tracking data	Ground fixed or mobile, airborne, space-based commercial.	Level 0, Level I Level II, Level III

Table 11.1

Comparison of Military Handbook-217 and Bellcore procedures

11.4.3 Example Military Handbook-217 Parts Count Versus Bellcore Method I

It may be helpful to provide an example that compares the Military Handbook-217 Parts Count and Bellcore Method I. This example is for a commercial electric clock. The failure rate prediction analysis is provided in Table 11.2, with the system reliability results given in Table 11.3.

11.4.4 Additional Methods and Techniques

Discussions on other methods and techniques are described below:

Rome Air Force Development Center (RADC) Toolkit

Many excellent references are available as guidance for calculating redundancy. The Rome Air Force Development Center Toolkit (see Reference 1) is a good reference because it uses language that is widely applied by the reliability practitioners. The Toolkit is also relatively inexpensive. It is a good reference for terminology and DfR methods. The Toolkit explains six redundancy conditions, and it provides the guidance for calculating the system failure rate based on any of the six situations. Three of the six disregard the effects of maintenance, and the other three include the effects of maintenance.

Bill of Space Material (BOM)	Qty	Military Handbook-217F, Notice 1 (GB, Parts Count)			Bellcore (GC, Method I)		
		Generic FR (FPMH)	pi Q	FR (FPMH)	Generic FR (FITs)	pi Q	FR (FITs)
Electrical Elements							
Electric motor, AC	1	1.60000	1	1.6000	500	2.5	1,250.0
Buzzer (piezo-electric crystal)	1	0.03200	2.1	0.0672	50	2.5	125.0
Switch, buzzer, on-off	1	0.00100	20	0.0200	15	2.5	37.5
Connector, AC power	1	0.01100	2	0.0220	10	2.5	25.0
Solder joints	6	0.00014	2	0.0017	5	2.5	75.0
Crimp joints	2	0.00026	2	0.0010	5	2.5	25.0
Electrical cord	1	note 1			note 1		
Mechanical Elements		note 2			note 2		
Gears	6	note 2			note 2		
Knobs	3	note 2			note 2		
Sweep hands	3	note 2			note 2		
Clock face	1	note 2			note 2		
Software Elements	0	note 2			note 2		
Operator Error Elements	1	note 2			note 2		
Total Failure Rate				1.7119			1,537.5

Note 1: Disregarded in Note 2. Not covered by this method.

Table 11.2
Comparison of Military Handbook-217 and Bellcore for a clock

Dimension	Military Handbook-217F	Bellcore
Failure rate	1.7119 (in FPMH)	1,537.5 (in FITs)
AFR in failures/year	0.01500	0.01347
MTBF in hours	584,139	650,407
MTBF in years	66.68	74.25

Table 11.3
Comparison of failure rates and reliability results for a clock

Estimating Complex Redundancy
Using Discrete Event Simulation Techniques

Sometimes an extremely complex redundancy situation occurs that cannot be evaluated using simple relationships. If the redundancy involves queuing conditions, simple relationships are not adequate. In very complex situations, discrete event simulation models are needed. A number of specific software tools are available to perform simulation predictions.

Estimating Temperature Acceleration Factors
Using the Arrhenius Relationship

The Arrhenius relationship is commonly used to estimate temperature acceleration factors. Details are described in Chapter 9. This relationship provides a convenient comparison of the effects of temperature for any device/failure mechanism with known activation energy. The Arrhenius relationship requires that you have two temperatures of interest and knowledge of the activation energy for failure mechanism. The model provides the acceleration (or deceleration) based on the different temperature conditions. The limitation is the knowledge about the activation energy and correlation to the failure mechanism.

Estimating Acceleration Factors
Using a Wide Range of Other Application Factors

There are other stress models besides Arrhenius. Other models that are a function of the stress and for a particular failure mechanism (metal fatigue, corrosion, electromigration, etc.) may provide more appropriate estimates on reliability (see Chapter 9). Often, the mechanical engineering group has insight into stress acceleration factors because they may have worked out finite element models on thermal and mechanical stress situations. The limitation for finite-element modeling is the analysis time to create and evaluate specific models for each specific situation of interest.

Estimating Reliability
Using Conditional Probabilities Methods

The statistics for combining probabilities is a well-known science, and an overview has been provided earlier in this chapter. System reliability often requires detailed probability analysis when serial models cannot be used. In this case, fundamental probability mathematics is required. Such modeling will most likely require expertise in this area, as software solutions will most likely be unavailable.

11.5 Common Problems

Any technical analysis can encounter problems while performing the analysis. Reliability predictive methods are no exception. This section describes a few of these problems.

A common question is "How do I use this prediction to improve the product?" While there are many answers to that question, the first question the analyst may ask is "Can I use this analysis for the intended purpose?" Obviously, this infers that a purpose guided the selection method before analysis started. The analysis should be reviewed first to ensure that it would have the precision needed for the application. This includes not only the analysis assumption, but also any realism factors to be used with the analysis. Once that hurdle has been cleared, the ability to use it should be easy.

Another common question concerns the library for the parts and the application stresses. It is customary to spend more time collecting the library and stress application information than it takes to input all of the information into one of the prediction tools. People who have not performed extensive evaluations often overlook the library importance.

Still another common question is related to the realism for the various evaluations. Unfortunately, many people have serious misunderstandings about the realism of various evaluation methods. None of the available industry-standard methods contains realism guidance or expectations. As a user, one should establish a realism approach for one's own applications and intended usage. Some experiences have been published in the public domain for guidance, but there is no substitute for developing one's own realism factors. For example, it is helpful to compare the original predictions to field results.

References
1. The Rome Laboratory Reliability Engineer's Toolkit, April 1993.

APPENDIX A

Tabulated *k* of *n* System Effective Failure Rates

For all units active with equal unit failure rates, and *k* out of *n* required for success as shown in Figure 11.A.1, the effective failure rate is given by [1]

$$\lambda_{eff\ k/n} = \frac{\lambda}{\displaystyle\sum_{i=k}^{n} \frac{1}{i}} \qquad (11.A.1)$$

Figure 11.A.1
k of n block diagram for Table 11.A.1

Results from this equation are tabulated below. As an exercise, verify the results in Example 11.4 using Equation 11.A.1 or the table below.

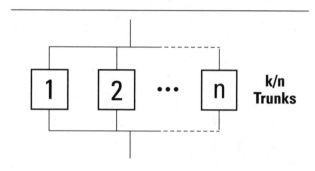

Table 11.A.1
Tabulated k of n system effective failure rate values

n	k	λeff		n	k	λeff
1	1	λ		5	4	$(60/27)\lambda$
2	1	$(2/3)\lambda$		5	5	5λ
2	2	2λ		6	1	$(60/147)\lambda$
3	1	$(6/11)\lambda$		6	2	$(60/87)\lambda$
3	2	$(6/5)\lambda$		6	3	$(60/57)\lambda$
3	3	3λ		6	4	$(60/37)\lambda$
4	1	$(12/25)\lambda$		6	5	$(60/11)\lambda$
4	2	$(12/13)\lambda$		6	6	6λ
4	3	$(12/7)\lambda$		7	1	$(140/363)\lambda$
4	4	4λ		7	2	$(140/223)\lambda$
5	1	$(60/137)\lambda$		7	3	$(140/153)\lambda$
5	2	$(60/77)\lambda$		7	4	$(420/319)\lambda$
5	3	$(60/47)\lambda$		7	5	$(210/107)\lambda$

APPENDIX B

Redundancy Equation with and without Repair

This appendix contains tabulated redundancy approximation given in Reference 1. These equations have the following notations: $\lambda_{k/n}$ is the effective failure rate of the redundant configuration where n of k units are required for success, n is the active online units, λ is the failure rate of an individual online unit (failures/hour), μ is the repair rate ($\mu = 1/Mct$ where Mct is the mean corrective maintenance time in hours), and P is the probability switching mechanism that will operate properly when needed ($P = 1$ with perfect switching).

1. When all units are actively online having equal unit failure rates with k out of n required operational for success, the effective failure rate with repair is

$$\lambda_{eff\,k/n} = \frac{n!(\lambda)^{n-k+1}}{(k-1)!(\mu)^{n-k}} \qquad (11.B.1)$$

and without repair is

$$\lambda_{eff\,k/n} = \frac{\lambda}{\displaystyle\sum_{i=k}^{n} \frac{1}{i}} \qquad (11.B.2)$$

2. When two active online units operate with different failure with one of two required for success, the effective failure rate with repair is

$$\lambda_{eff\,1/2} = \frac{\lambda_A \lambda_B [(\mu_A + \mu_B) + (\lambda_A + \lambda_B)]}{\mu_A \mu_B + (\mu_A + \mu_B)(\lambda_A + \lambda_B)} \qquad (11.B.3)$$

and without repair is

$$\lambda_{eff\,1/2} = \frac{\lambda_A^2 \lambda_B + \lambda_A \lambda_B^2}{\lambda_A^2 + \lambda_B^2 + \lambda_A \lambda_B} \qquad (11.B.4)$$

3. When one standby offline unit with n active online units required operating for success (with offline spare assumed to have a failure rate of zero), and online units have equal failure rates, then the effective failure rate with repair is

$$\lambda_{eff\,n/n+1} = \frac{n[n\lambda + (1-P)\mu]\lambda}{\mu + n(P+1)\lambda} \qquad (11.B.5)$$

and without repair is

$$\lambda_{eff\,n/n+1} = \frac{n\lambda}{P+1} \qquad (11.B.6)$$

APPENDIX C

Availability

The basic mathematical definition of availability is

$$\text{Availability} = A = \frac{\text{Up Time}}{\text{Total Time}} = \frac{\text{Up Time}}{\text{Up Time} + \text{Down Time}} \qquad (11.C.1)$$

Actual assessment can involve substituting the time that comes from various forms of this basic equation. Thus, different combinations of elements combine to formulate different definitions of availability. Two different types of availability are described.

Inherent Availability (Ai)

Inherent availability is used when a system's availability is defined with respect only to operating time and corrective maintenance, and is

$$\text{Ai} = \frac{\text{MTBF}}{\text{MTBF} + \text{MTTR}} \qquad (11.C.2)$$

where *MTTR* is the Mean Time To Repair an item and *MTBF* is the Mean Time Between Failure. Under this definition, other time periods are ignored, such as standby, scheduled delay times, preventative maintenance, as well as administrative and logistic down time. Inherent availability is useful in determining basic system operational characteristics. However, it provides a very poor estimate of the true system's potential, because it provides no indication of the time needed to obtain required field support.

Operational Availability (Ao)

Operational availability covers all segments of time that the equipment is intended to be operational. Here, up time now includes operating time plus non-operating (standby) time (when equipment is assumed to be operable). Down time is expanded to include preventive and corrective maintenance and associated administrative and logistic lead time. All are measured in clock time.

$$\text{Ao} = \frac{\text{OT} + \text{ST}}{\text{OT} + \text{ST} + \text{TPM} + \text{TCM} + \text{ALDT}} \qquad (11.C.3)$$

Here *OT* is the operating time in use, *ST* is the standby time, *TPM* is the total preventive (scheduled) maintenance time, *TCM* is the total corrective (unscheduled) maintenance time, and *ALDT* is the administrative and logistic down time (delay-down time with no maintenance time). This definition is intended to be a realistic measure of equipment availability. A simpler common expression that is often used for Ao is

$$\text{Ao} = \frac{\text{MTBM}}{\text{MTBM} + \text{MDT}} \qquad (11.C.4)$$

where *MTBM* is the Mean Time Between Maintenance actions and *MDT* is the Mean Down Time.

Availability and Probability

Availability is essentially a probability number describing the probability to be available. From the way it is defined (up time/{up time + down time}), availability is a number between 0 and 1. Therefore, availability can be treated mathematically, similar to the way in which reliability, R, is described. For example, if the availability of subsystems A, B, and C is 0.9, 0.95, and 0.8, respectively, then the availability of these subsystems in series is $(0.9)(0.95)(0.8) = 0.684$. Thus, the serial system is available 68.4 percent of the time. Such treatment can be important in simplifying availability estimates.

CHAPTER 12

Failure Modes and Effects Analysis

12.1 Failure Modes and Effects Analysis

A Failure Modes and Effects Analysis (FMEA) helps to evaluate numerous aspects of a product. An FMEA identifies many specified and unspecified customer requirements related to product design, its use, how failures may occur, the severity of such failures, and the probability of the failure occurring. With these identified, a team can focus on the design process and the major issues facing the product in its potential use environment for the customer. The FMEA approach recommended in this chapter is a team evaluation that considers a product or process failure mode (the loss of product function) and then describes the expected failure effect at the next higher level (the effect to the customer). The first step in a good FMEA is to define the purpose(s) or expectations of this evaluation. Some common purposes are: collect the voice of the customer, improve first-pass success of projects, address failure modes early, perform design trade-off studies, select design controls, and provide risk control.

The FMEA team identifies the product functions and then selects the failure modes. The general guidance is to select the major functions and to disregard those that are considered minor in terms of the purposes defined for the project. After considering the major functions, other functions can be selected, if needed, to support the project. Failure effects should be described in functional terms that may be understood by a customer. FMEA activities are planned early in the development process of new products and new manufacturing processes in order to meet customer needs on time.

12.2 FMEA Goal and Vision

A key FMEA goal is the early identification of potential issues that can affect the customer. Early identification is necessary to achieve first-pass successful development of a new product or process. Early evaluation helps to identify unspecified customer needs, since it is performed from a customer-use perspective. This reduces project risks and improves reliability, yield, and profitability. The FMEA program described here is an integrated element of the overall Design for Reliability (DfR) strategy within the stage gate process. The basic FMEA objective is "Using this evaluation process to continuously improve the successful development of new products and processes." An

Figure 12.1
FMEAs ensure that a design reaches its full potential

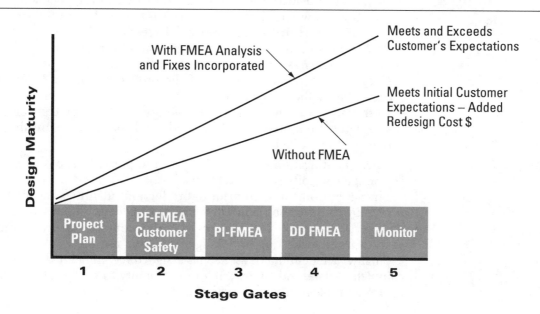

FMEA performed over the stage gate process ensures that a design reaches its full potential (see Figure 12.1).

12.3 FMEA Concepts

FMEA is a structural method to study a design or process that seeks to anticipate and minimize unwanted performance or unexpected failure. The main objective is to improve first-pass success through the early identification of issues that can affect customer use. FMEA asks the question, "What can go wrong?" even if the product meets specification. FMEA methodology provides a structural method with resulting documentation to aid in meeting project objectives. The name FMEA can be broken down in related terms of:

- *Failure Mode*
  The manner in which a part or assembly could potentially fail to meet its requirements or fail to function. (Example: Telephone – ringer fails.)
- *Effects*
  The potential nonconformance stated in terms of the next assembly or system performance. (What will the customer notice?)
  (Example: Ringing subsystem inoperable; customer misses calls.)
- *Cause*
  The potential reason(s) behind a failure mode, usually stated as an indication of a specific design or process weakness.
  (Example: Wire overheats and opens – improper wire gauge selection.)
- *Analysis & RPN*
  By performing an FMEA, failure modes will be anticipated. Risks can be determined and assessed for the customer/product. Recommended actions can be associated to neutralize the risk to an acceptable level. (Example: Wire size too risky; add-on cost required for heavier wire size.)
  These risks are quantified using a Risk Priority Number (RPN).

To help assess the risk associated with each failure mode, the analysis uses a rating system to quantify the failure mode's Severity, probability of Occurrence, and Detection. The product of these numbers yields the failure mode's Risk Priority Number. This number provides a rank for priorities associated among failure modes. An overview of the key factor involved in the RPN value is provided below and with more detail in Appendix A.

- The RPN is a way to quantify risk. This is the product of three numbers:
  RPN = Severity × Occurrence × Detection
- *Severity* = A rating (1–10) of the seriousness of the failure mode effect on the next higher assembly, the system, or the user.
- *Occurrence* = A rating (1–10) of the probability of the failure mode to occur during the design cycle.
- *Detection* = A rating (1–10) of the ability of the design control to identify and detect the potential failure mode and its cause before the design is released to production.

Once quantification is assessed, the next important issue is to establish the key design control(s) (see *Detection* above).

The design control (used prior to production) identifies key method(s) for preventing the cause of failure from occurring, detecting the cause and initiating a corrective action. Examples of design controls are design verification testing, prototype testing, and design validation testing.

Finally, each item that is assessed is assigned recommended actions, a target completion date, and engineering responsibility to complete the actions. An FMEA example is provided in Section 12.6.

12.4 Types of FMEA Evaluations

Many types of FMEA evaluations are currently in use. In this chapter, four major FMEA types are defined and recommended to support a range of projects and development efforts in preparing products that best support the customer's needs. The four major FMEA types fall under the category of a design or process FMEA (see Figure 12.2).

A design evaluation looks at the product while a process evaluation looks at the product manufacturing process steps. A traditional design FMEA starts with the identification of a possible failure. Then the evaluation describes the failure effects at the customer level. FMEA is team-oriented to help obtain numerous inputs from various team members who are familiar with the product and the customer's needs. Figure 12.3 shows the typical FMEA activities that occur in various stage gate periods.

Figure 12.2
FMEA families and types

Types	Design	Process
Families	PF-FMEA	PP-FMEA
	PI-FMEA	
	DD-FMEA	

Key FMEA program definitions follow in the same sequence shown in Figure 12.3.

FMEA Project Planning: During the Idea Phase of the stage gate process, the DfR tasks are defined. One of the DfR considerations is planning the FMEA tasks. This is the point where it is decided what value is expected from an FMEA evaluation and to make a commitment to either perform it or to omit it from the other scheduled DfR project tasks. This depends on the customer's needs.

Figure 12.3
FMEA stage gate activities

The Product Function Failure Modes and Effects Analysis (PF-FMEA) is used to evaluate the effects at the customer level following a failure of each of the top-level product functions. This is a strong tool for understanding customer needs as each function is examined with respect to its intended use. This is commonly performed during the evaluation stage of the stage gate development process. The PF-FMEA is the simplest of the FMEA evaluation options, and it can be performed at the earliest point in the development cycle.

The Product Interface Failure Modes and Effects Analysis (PI-FMEA) is used to evaluate the effects at the customer level following a failure of each of the top-level product input and output connections. It helps to understand and examine closely customer-specified interface requirements. A PI-FMEA is usually performed during the development stage in the stage gate development process. The PI-FMEA is also simple, but the product must be well formulated before the analysis can be performed since it considers the failure modes to be the failure of each input and output connection.

Stage Gate	FMEA Activity Expected
Idea Stage	FMEA Project Planning
Evaluate	Product Function FMEA (PF-FMEA)
Development	Product Interface FMEA (PI-FMEA)
Transition	Detail Design FMEA (DD-FMEA)
Production	Product Process FMEA (PP-FMEA)
	Updates if necessary

The Detail Design Failure Modes and Effects Analysis (DD-FMEA) is the most common FMEA performed throughout industry. It is used to evaluate the effects at the customer level following a failure of each product part. This detailed analysis examines design aspects that could affect customer requirements at the technical level. This is not expected to be performed on many projects, but when it is performed, it is done during the transition stage in the stage gate process. The DD-FMEA is the most complex analysis. If it is performed, the development must be substantially finished.

The Product Process Failure Modes and Effects Analysis (PP-FMEA) is used to evaluate effects at the customer level following a failure in each of the product manufacturing process steps. This is an important evaluation as it closely examines potential manufacturing problems that could have an impact on meeting customer requirements. This analysis is performed during the production stage of the stage gate process.

12.5 Objectives

Some typical FMEA program objectives are as given in Figure 12.4. Each of the five objectives is discussed in the following paragraphs.

Figure 12.4
FMEA objectives

✓ Support FMEA program vision.

✓ Increase activities with customer focus.

✓ Use a team approach.

✓ Support continuous improvement.

✓ Optimize lessons learned.

✓ Use concurrent engineering best practices.

Support FMEA program vision. Today's leading industries have DfR guidance that includes FMEA. The stage gate development process contains a requirement to plan FMEA tasks during the idea stage. Stage gate recognizes that there is a wide range of risks and costs to consider in selecting any of the development tasks. The DfR guidance has been formulated consistently with the types of projects each business unit is expected to manage.

Increase activities with customer focus. In the traditional implementation approach, the FMEA team discusses failures and failure effects. In addition, customer focus assures a heightened awareness that customer requirements will be met. A persistent question is "Will the product functionally serve the customer's needs, and has reliability been designed in it?" When a product fails, a customer is most acutely aware of the product and the loss of the desired functionality.

Use a team approach. A third FMEA project objective is to perform FMEA evaluations in a team-oriented manner to gain more technical expertise for the evaluation. By focusing on customer requirements, a team-oriented approach provides numerous inputs to help assure that all the customer needs are met. The use of multifunctional teams is very effective during the development process.

Support continuous improvement. Documenting and applying consistent methods are key benefits observed by many ISO implementations. This structure provides a foundation to lower costs and fosters continuous growth.

Optimize lessons learned. Consistent methods automatically help improve the feedback from project to project. The documentation of conclusions also helps with learning as a project progresses. Some industry leaders have reported that such FMEA evaluations are being effectively used for orientation to new project team members. FMEA studies are also useful to the manufacturing staff and customer service representatives when they join an existing development team.

Use concurrent engineering best practices. Best practices examine all development tasks and identify tasks that are a series or parallel in the development cycle. In an efficient concurrent engineering schedule, most of the required tasks are performed in a parallel configuration rather than serially. The tasks that constrain the schedule are critical path serial tasks. The person selected as the FMEA team leader should be someone who is not performing the majority of the critical path efforts during that stage of the project. In fact, it is acceptable to have an FMEA team member not attend the FMEA team meeting. An FMEA team member can independently review the documented team consensus and add opinions before a final report is released.

12.6 An FMEA Example

An effective way to understand FMEA is through an example. Figure 12.5 shows results for an electrical PI-FMEA. Some of the FMEA findings are discussed below.

- This is a PI-FMEA that evaluates the loss of each individual input and output. Since this is a functional FMEA, the first column is Item Function. Each pin is evaluated individually. The example covers Pin #1 through Pin #6. The function of Pin #1 is shown as an electrical ground.
- In the Potential Failure Mode column, selected are some open and shorted potential failure modes to evaluate.
- In Potential Effect(s) of Failure at the customer level, the example shows "No effect for 1 of 3" since three parallel ground paths exist and any single

Figure 12.5

FMEA example form

											Action Results				
	Failure Mode and Effect Analysis														

System: XXX — (PI-FMEA) — FMEA Number: XXX
Subsystem: XXX — Page: 1 of 2
Component: XXX — Prepared by: XXX
Model Year(s)/Vehicle(s): XXX — FMEA Date (Orig.):
Core Team: — FMEA Update:

Item Function	Potential Failure Mode	Potential Effect(s) of Failure	SEV	CLASS	Potential Cause(s)/ Mechanism(s) of Failure	OCCUR	Current Design Controls	DETEC	RPN	Recommended Action(s)	Responsibility & Target Completion Date	Actions Taken	SEV	OCCUR	DETEC	RPN
Electrical	open	No effect	1	c	solder joints or	1	Note 1	6	6	N/A						
Pin #1	open	for 1 of 3	1		metalization	1		6	6	N/A						
Electrical																
Ground																
Electrical	open	Loss of RF	10	c	Hot R3 or	6		3	180	Note 2						
Pin #2	open	generation	10		solder joints or	1	Note 1	6	60	N/A						
Electrical	open		10		metalization or	1		6	60	N/A						
V osc	shorted		10		FET	4		3	120	Note 3						
Electrical	open	Degradation	10	c	solder joints or	1	Note 1	6	60	N/A						
Pin #3	open	of range	10		metalization or	1		6	60	N/A						
Electrical	shorted	evaluation	10		varactor	2		2	40	N/A						
V tuning																
Electrical	open	No effect	1	c	solder joints or	1	Note 1	6	6	N/A						
Pin #4	open	for 1 of 3	1		metalization	1		6	6	N/A						
Electrical																
Ground																
Electrical	shorted	Degradation of	8	c	Mixer diode or	3		2	48	N/A						
Pin #5	open	performance	8		solder joints or	1	Note 1	6	48	N/A						
Electrical	open		8		metalization	1		6	48	N/A						
+IF																
Electrical	shorted	Degradation of	8	c	Mixer diode or	3		2	48	N/A						
Pin #6	open	performance	8		solder joints or	1	Note 1	6	48	N/A						
Electrical	open		8		metalization	1		6	48	N/A						
−IF																

ground pin being open of the three will cause no problem at the customer level. However, in the open Pin #2 example, the Potential Effect(s) of Failure shows "Loss of RF generation" at the customer level. This happens because Pin #2 provides the electrical voltage for the oscillator, and loss of that voltage interrupts the RF generation, interrupting customer usage.

- The information in the Severity (SEV) code column is assigned using FMEA key criteria (see Appendix). This example has a range of Severity codes from 1 to 10. The Severity range is 1 for a small severity and 10 for a large severity. A Severity of 10 is identified for several of the failure modes in the example. A 10 corresponds to potentially killing a customer and requires a warning. This is a serious safety issue. Safety issues are common findings in the early FMEA stage, before mitigation can be implemented.
- The classification (CLASS) column may be used to classify any special product characteristic that may require additional process controls. The criteria for assigning CLASS are in the Appendix. The "c" listed in this example is to identify a sensitivity for ESD.
- The list of "Potential Cause(s)/Mechanism(s) of Failure" shows the failure mechanisms that caused the listed failure mode.
- The list of Occurrence probability (OCCUR) shows the rating the team assigned to rank the probability of Occurrence. A low number means a low probability. The guidance for assigning "OCCUR" is also in the Appendix.
- "Current Design Controls" describes the current design controls that are intended to control the frequency. In the example, "Note 1" indicates that the space on the form was too small for the required information.
- DETEC is the rating the team assigned to show the Detection capability. The guidance for assigning DETEC is in the Appendix.
- A Risk Priority Number (RPN) is a simple product of the Severity code (SEV), Occurrence probability (OCCUR), and Detection capability (DETEC). The higher the value of RPN for any failure mode, the more importance it should get for corrective action. The corrective actions usually have a direct impact on meeting and/or exceeding customer requirements and, thus, customer satisfaction.
- "Recommended Action(s)" is the location to document the planned actions to resolve identified weaknesses. In the example, "Note 2" and "Note 3" were used since the space on the form was insufficient for the details.

12.7 Implementation Methods

This section provides common help on basic methods for implementing an FMEA. There are three areas of interest in this section: common FMEA team meeting agenda, faster methods for implementing a team-oriented design FMEA, and FMEA process flow and responsibilities.

Common FMEA Team Meeting Agenda

An overview of the common steps performed in a team-oriented FMEA are provided below.

1. FMEA Team Leader performs the preparatory efforts (i.e., collecting the requirements and past similar evaluations, selecting and inviting the team, and preparing the review forms).
2. During the FMEA team meeting, the FMEA Team Leader either facilitates the meeting or assigns a person to do the facilitating.
3. Discuss any customer requirements.

4. Discuss and select the FMEA evaluation purpose.

5. Discuss and select the analysis method.

6. Discuss and select the product baseline.

7. Discuss and select the customer view.

8. Discuss and select the failure modes.

9. Discuss and select the failure effects.

10. Assign severity ratings.

11. Review RPN values, if applicable.

12. Discuss actions and responsibilities as applicable.

13. Discuss FMEA evaluation updating criteria.

14. Discuss FMEA follow-up meeting needs.

15. Conduct FMEA follow-up meetings, if needed.

16. Prepare, review and approve FMEA report.

17. Perform the FMEA Project Evaluation.

Faster Methods for Implementing a Team-Oriented Design FMEA

Design team-oriented FMEAs can take a lot of time. This represents many man-hours. Long meetings can become unproductive after a while. Therefore, it is important to extract key information early. To accomplish this task, the following approach may prove more efficient:

1. Follow procedure above for a common FMEA team meeting agenda.

2. Review potential environmental failure modes first to help find major problems quickly due to: temperature cycles – solder joint failure, TC mismatches, bond problems; temperature – high junction temperatures, diffusion, intermetallics; voltage overstress – voltage surge protection, shorts, open, high E fields; ESD exposure – input and output pins, ESD protection required; shock and vibration – large components, board flexures, position of large components on the board; humidity related – corrosion, and dendrites/Ag epoxy, water, seal. This approach is not well known but can provide significant time-saving. Here, one looks at environmental stress effects to identify what failure modes can occur as a result of a particular applied stress. These focus quickly on a subcategory of stress-related failure

Figure 12.6
Design FMEA worksheets

System: Subsystem: Design Responsibility: Core Team
Item:
Function:
Potential Failure Mode:
Potential Effects of Failure Mode:
Possible Causes, Failure Mechanism:
Current Design Controls:
Recommended Action:
Severity   Class   Occurrence   Detection   R.P.N.
...etc.

modes that help the teams focus on other components that could have similar problems. This triggers a team thought process that is highly efficient in FMEA. For example, failure modes related to temperature cycle stress quickly identify temperature coefficient problems throughout the product. Temperature expansion-contraction failure mechanisms provide a global team view assisting the team to identify numerous common problems, causes, design controls, ratings, recommended actions, etc. The results yield a highly efficient team approach.

3. Perform a part-by-part review. Ask the question "What are the potential failure modes for each part?" This is the common approach. However, when integrated after Step 2, it provides an opportunity to look for any other types of problems that may have been overlooked in Step 2.

4. Avoid spending time listing superficial failure modes. Often the team will be sidetracked early on superficial details that can burden the team's time, resulting in a loss of quality time on key issues.

5. Use the suggested FMEA worksheet shown in Fig. 12.6. After the meeting, organize the results on a proper FMEA form.

FMEA Process Flow and Responsibilities

An overview of this FMEA process flow is as follows.

1. Program Managers need to be informed about the recommended FMEA program.

2. Program Managers must establish the FMEA project needs by assuring the customer's related requirements are defined for FMEA reporting.

3. An FMEA Team Leader is assigned and advised what purpose is being considered.

4. The Team Leader's responsibilities include a review of past similar analyses, a review of related customer requirements, project planning, and implementation of the planned effort including performing an FMEA project evaluation when it is completed.

5. The Program Manager, FMEA Team Leader, and FMEA Administrator share a responsibility to identify further changes to the FMEA process.

6. When an FMEA defines an action plan for improvement, the responsible members of the FMEA team implement the action plan.

7. Management continues to provide the enabling policies, procedures, and staffing.

APPENDIX A

Guide to Assigning FMEA Key Criteria

The Risk Priority Number (RPN) is calculated as the product of Severity (SEV), Occurrence probability (OCCUR), and Detection capability (DETEC) numbers. This key FMEA index helps to quantify risks. The RPN is defined below. Early FMEA evaluations may not have all of these aspects defined at an early development stage. Severity ratings are expected to be completed during all FMEA projects.

The rating system should also include Classification (CLASS) to identify any special controls needed during the product manufacturing, assembly, or testing.

Risk Priority Number

Risk Priority Number (RPN) is a quantitative measure to evaluate and assess the failure mode (see Table 12.A.1). This characteristic is automatically calculated using the criteria of the three subelements. The RPN is the result of three other measurements:

- Severity = SEV
- Occurrence probability = OCCUR
- Detection capability = DETEC

Table 12.A.1
RPN interpretation

Rank	Guideline
$1 < RPN < 18$	Minor product and/or business risk.
$18 < RPN < 64$	Moderate risk. This requires selective product validation and evaluation of design and/or process characterization to reduce the RPN measure.
$64 < RPN$	Major risk. Requires extensive design and/or process revisions to reduce the RPN measure.

Risk Priority Number (RPN) Calculation and Significance

Risk Priority Number (RPN) is a measure based on the product of Severity, Occurrence probability, and Detection capability.

$$RPN = (SEV) \times (OCCUR) \times (DETEC)$$

Table 12.A.2
Severity (SEV) ratings

Rating	Guideline	Rank
Very High	Indicates a potential failure mode that could cause death (9 with warning, 10 without warning).	10 9
High	High customer dissatisfaction due to the nature of the failure, such as a major system (e.g., automobile engine) function being inoperative.	8
High to Moderate	Can also be an inoperable convenience system (e.g., air-conditioning system). Do not involve safety aspects.	7
Moderate	Failure causes some customer dissatisfaction.	6
Moderate to Low	Customer is made uncomfortable or is annoyed by the failure.	5
Low	Customer will notice some subsystem or vehicle performance deterioration.	4
Low to Minor	The nature of failure causes only slight annoyance. The customer will probably only notice a slight deterioration of the performance.	3
Minor	Unreasonable to expect that the minor nature of this failure would cause any real effect on the system performance.	2
Very Minor	Most customers would probably not even notice the failure.	1

When SEV, OCCUR, and DETEC are assigned values described in this Appendix, the RPN number can then be interpreted as shown in Table 12.A.1.

The ranking system for each key rating criteria is based on a scale from 1 to 10 (see Table 12.A.2).

Severity (SEV)

If a Severity is rated as a 10, the development staff should aggressively try to solve or mitigate the severity before products are delivered (see Table 12.A.2). Such early identification during product development will improve delivered products.

Closure can be by:

- Corrective action to reduce the Severity rating.
- Mitigation to reduce the Severity rating.
- Notification to partner/customer so a system-level solution is established to reduce the Severity.

Occurrence Probability (OCCUR)

The FMEA team may alter this characteristic. It is not necessary for the likely failure to be in the same units as listed. The failure rate can be expressed as failures per million hours or as a probability (see Table 12.A.3).

Detection Capability (DETEC)

The recommended detection criterion is based on the ability of a design maturity test to detect a particular failure mode. This criterion is useful if the purpose is to evaluate the intended testing program. For example, the Department of Defense FMEA standard uses the system fault detection criteria as the detection rating. A common rating system is shown in Table 12.A.4. The team may alter these detection criteria as needed.

Classification (CLASS)

This column may be used to classify any special product characteristics for components, subsystems, or systems that require additional process controls. Any item deemed to require special process controls should be identified on an FMEA form with the appropriate character or symbol in the classification column and should be addressed in the recommended action column. Each item identified above in a design-type FMEA should have the special process controls identified in the process-type FMEA.

Table 12.A.3
Occurrence and other expressions for the failure rate

Failure Rate	Likelihood of Failure	Ranking	Occurrence per Unit Time	
Very High	Failure is almost inevitable.	10	1 in 2	(50%)
		9	1 in 3	(33%)
High	Repeated failures.	8	1 in 8	(12.5%)
		7	1 in 20	(5%)
Moderate	Occasional failures.	6	1 in 80	(1.25%)
		5	1 in 400	(0.25%)
		4	1 in 2,000	(0.05%)
Low	Relatively few failures.	3	1 in 1,500	(666 PPM)
		2	1 in 150,000	(6.66 PPM)
Remote	Failure is unlikely.	1	1 in 1,500,000	(0.66 PPM)

Rating	Guideline	Rank
Certainty of Nondetection	Screening cannot detect a potential failure mechanism, or there is no screen.	10
Very Low	Screening probably will not detect a potential failure mechanism.	9
Low	Screening not likely to detect a potential failure mechanism.	8 7
Moderate	Screening may detect a potential failure mechanism.	6 5
High	Screening has a good chance of detecting a potential failure mechanism.	4 3
Very High	Screening will almost certainly detect a potential failure mechanism.	2 1

Table 12.A.4
*Failure mode
detection ratings*

APPENDIX B

FMEA Forms

Figure 12.B.1
Potential Failure
Mode Effects Analysis
(Design FMEA form)

Item/ Function	Potential Failure Mode	Potential Effect(s) of Failure	SEV	Potential Cause(s)/ Mechanism(s) of Failure	OCCUR	Current Design Controls	DETEC	RPN	Recommended Action(s)	Responsibility & Target Completion Date	Action Results				
											Actions Taken	SEV	OCCUR	DETEC	RPN

Figure 12.B.2
Potential Failure
Mode Effects Analysis
(Process FMEA form)

Process/ Function	Potential Failure Mode	Potential Effect(s) of Failure	SEV	Potential Cause(s)/ Mechanism(s) of Failure	OCCUR	Current Process Controls	DETEC	RPN	Recommended Action(s)	Responsibility & Target Completion Date	Action Results				
											Actions Taken	SEV	OCCUR	DETEC	RPN

CHAPTER 13

Evaluating Product Risks

13.1 Introduction

This chapter addresses the issues of technical risk management and can be used as guidance for all technical areas. Risk management applies to all new product development. Common technical risk areas include performance, producibility, production, scheduling, resources, and so forth. Risk varies depending on whether customer requirements match technology performance capability predictions, if field experience is available on analogous assemblies, if the technology is revolutionary or evolutionary, if the application is new, if the intended use environment is harsh and different from previous field experience, and so forth. Risks are often assessed in categories. A technology management risk matrix is often used in industry (see Figure 13.1).

As Figure 13.1 shows, revolutionary technologies carry a higher risk. For example, when the first airplanes were developed in the early 1900s, flying these early machines often resulted in injury or death. Now that flying is a mature technology, the risks of flying are very low. Evolutionary changes to the aircraft having similar applications today carry low risks since the technology is mature.

Figure 13.1
The management technology matrix

	Evolutionary	Revolutionary
Same Application	Category I (Low Risk)	Category III (High Risk)
New Application	Category II (Moderate Risk)	Category IV (Very High Risk)

Placement within the matrix determines the degree and aggressiveness of the management program.

13.2 Goals of a Risk Program

The goal of a risk management program is to make correct decisions at key points in the program. Technology risk management is essential to the success of any development program. Risk issues and their consequences concern everyone involved with a program's success. The larger and the more undeveloped a technical program, the more important it is to manage risks. In the case of a reasonably large and/or complex program, many technical details can impact the system. This chapter is designed to help mitigate risks. To help in

Table 13.1
Applicable sections of this reliability manual

Applicable Chapters of This Reliability Manual	Category I Low Risk	Category II Moderate Risk	Category III High Risk	Category IV Very High Risk
1. Reliability Science/ Design for Reliability			√	√
2. Understanding Customer Requirements	√	√	√	√
3. Design Assessment Reliability Testing			√	√
4. Design Maturity Testing	√	√	√	√
5. Screening and Monitoring			√	√
6. Process Reliability			√	√
11. Reliability Predictive Modeling		√	√	√
12. Failure Modes and Effects Analysis		√	√	√

the use of this chapter, rank your technology according to the categories in Figure 13.1. Refer to the applicable chapters of this manual for associated reliability items in Table 13.1 that match your technology rankings.

Table 13.1 indicates applicable chapters to aid in mitigating your risk. Even low-risk issues can become costly. Therefore, if you have a low-risk product, you may still wish to refer to the details below. The benefits of full risk management are shown in Figure 13.2.

Figure 13.2
Benefits of risk management

✓ Expose high-risk areas and critical parameters early in the program.

✓ Help direct resources by providing insight into potential consequences to allow for informed program decision-making.

✓ Identify and track actions to minimize risk and ensure resolution of key issues.

✓ Provide information to help Program Managers select an appropriate subsystem/component.

✓ Identify areas of risk that are potentially most harmful.

✓ Minimize liability risk.

Since component and subsystem risks are magnified at the system level, it is important that program management becomes aware of issues early in the program. All potential risk areas require identification and risk handling. Management can then direct resources to prioritized risk areas and conserve valuable time and expenses. These benefits are best realized when technical risk issues can be properly identified, assessed, quantified, and finally handled both at the system and the subsystem level.

13.3 Managing Risks for Your Program

The risk management process can be set up so it is reasonably formal, systematic, and applied in a disciplined manner. Figure 13.3 shows the classical systematic approach to risk management. A systematic approach will ensure that each element of risk planning, risk assessment, risk analysis, and risk handling is managed. Each element is described in this chapter. The easiest way to qualitatively manage a product's risk is to review the elements in Figure 13.3 and appropriately identify, in your Work Breakdown Structure (WBS), key events that are *potential risk factors*. Every program is different, and unfortunately, no magic approach can guarantee that risks are minimized. *Remember, the goal of a risk management program is to make correct decisions at key points in the program.* Decision management is risk management, and decisions should be

Figure 13.3
Technology maturation management

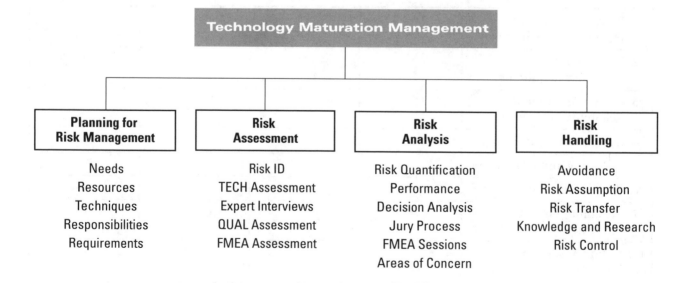

Technology Maturation Management			
Planning for Risk Management	**Risk Assessment**	**Risk Analysis**	**Risk Handling**
Needs	Risk ID	Risk Quantification	Avoidance
Resources	TECH Assessment	Performance	Risk Assumption
Techniques	Expert Interviews	Decision Analysis	Risk Transfer
Responsibilities	QUAL Assessment	Jury Process	Knowledge and Research
Requirements	FMEA Assessment	FMEA Sessions	Risk Control
		Areas of Concern	

based on information. The probability of making a correct decision is higher when correct information is obtained and made available in a timely manner. Following simple risk-management guidelines can save a program dollars. Follow the guidelines provided in each element to ensure that you are able to make the correct decision in a timely manner.

FMEA – A Reliability Method for Evaluating Product Risk

Although risk, as described here, applies to all facets of a project, a good example of how risk is managed in reliability is in FMEA. FMEAs can be viewed as one type of risk management (see Chapter 12). In an FMEA, all phases of risk management shown in Figure 13.3 are performed, including risk management planning, assessment, analysis, and handling. The progression provided in Figure 13.5 is followed in a team-oriented FMEA, where a brainstorming session is held to perform the evaluation, identify failure-mode issues, and quantify risks in terms of Severity, Occurrence, Detectability, and an RPN number. Finally, design controls and recommended actions are detailed to help mitigate and handle risks.

Figure 13.4
Work breakdown structure

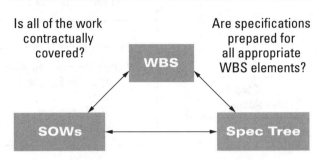

Is all of the work contractually covered?

Are specifications prepared for all appropriate WBS elements?

Are specifications properly included in all SOW(s)?

Does the WBS represent what is to be done?

Are all elements of the project WBS present?

Is it clear who owns what elements?

13.4 Four Steps to Risk Management

Figure 13.3 illustrates the elements of the risk-management process. Working through these elements in steps can perform risk management. Figure 13.5 shows the process. Starting with risk planning (Step 1), a brainstorming session should be held to overview the WBS or the project's overview. The purpose of the session is to identify concerns with such areas as meeting a project's needs, its resources, schedule, performance, reliability, and so forth.

All the areas of concern should be formally categorized into risk assessment (Step 2). This helps to organize and plan appropriately while identifying departmental responsibilities. At this point, each department can further detail the risks involved in its area and offer feedback into the program plan. The decision process can start. Decisions should be based on information. This is the point of risk management when decisions are made on program needs, gaps, and further information and testing that must be performed to more fully understand risks and make intelligent decisions.

Risk should be quantified when possible. This is part of the risk analysis noted in Step 3. Key to the program's success is the ability of the technology to meet or exceed customer performance expectations. Performance targets are often well defined in customer specifications. If the technology is revolutionary and part of a new application, expert opinion should be used when data are not available.

Estimates should be made as early as possible in the program as to whether the unit can meet performance, reliability, and other requirements. If the unit has only a 90% chance of meeting an important requirement and this jeopardizes the whole program, the Program Manager should be aware of the risks. For example, if the program is worth $10 million, the financial risk is 90% of this, or $9 million.

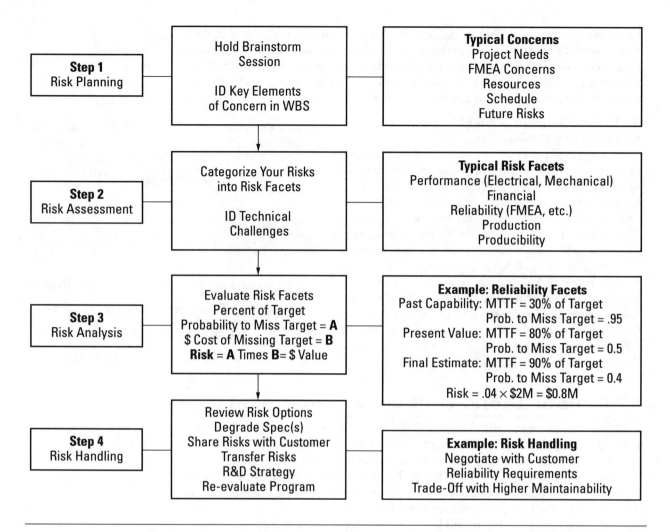

Figure 13.5
Four steps in risk management

At this point, risk handling is required (Step 4). Management needs to assess the options, such as sharing the risk with the program's customer by negotiating specifications, contractual agreements, trading off for tighter specification in other areas, and so forth. A watchlist is also required in risk handling. This list serves to identify scheduling problems, assess previously identified risks, update performance capabilities, and so forth. This chapter further details these steps in the risk-management process.

13.5 Guidelines for Risk Planning (Step 1)

This first step in a technology maturation program plan should include a risk-management plan. To plan for risk management, the five major areas identified in Figure 13.3 (under its block) need to be addressed. These are described in compact notation in Table 13.2.

Areas of Concern	Description	Guidelines
Needs	Coordinating program needs	Needs include personnel, appropriate teams, and suppliers. Eliminate and minimize the effects of undesirable occurrences.
Resources	Identifying resource problems	Establish time, money, and/or engineering reserves to cover risks that cannot be avoided.
Techniques	Systematic approach	Providing a formal and systematic risk-management approach is integral to the program's success and key to decision-making.
Responsibilities	Assigning and ensuring responsibilities	Document all risk for accountability so that appropriate engineering staff closely watches identified risk areas.
Requirements	Identifying future risk needs	Ensure important items undergo complete risk assessment, analysis, and handling as part of risk management.

Using the program's WBS/customer specification, work through the table to identify the areas of concern. For each area, if necessary, schedule a separate brainstorming session with area experts to both help plan and start to perform risk assessment.

Table 13.2
Guidelines for risk management

13.6 Guidelines for Risk Assessment (Step 2)

The second step in technology risk management is to assess risk (see Table 13.3). Risk needs should be identified and categorized into appropriate risk facets first so that responsibilities can be assigned to further clarify the risk category.

Common risk facets such as performance, reliability, and resources are shown in Figure 13.6. Each risk has associated challenges and tasks related to reducing and eliminating the risk. This assessment is initially qualitative and should be evaluated and identified as soon as possible.

Figure 13.6
Technology risk facets

After a gross survey of the challenges, the assessment needs to be refined. This should include expert opinions from experienced individuals. The decision process goes from being qualitative to quantitative, to make assessments more accurate regarding problems. At this point, risk analysis should be performed.

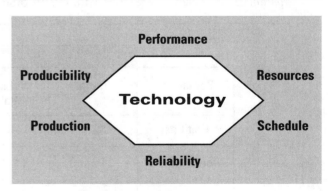

Areas of Concern	Description	Guidelines
Risk Identification/ Facets	ID technology risk and categorize into appropriate risk facets (Figure 13.6)	Identify risk and understand its relationship to the technology. Establish an organized approach to categorizing risk into appropriate facets.
Technology Assessment	Identify technical challenges that may fail	Provide an assessment of risk associated with evolving a new design, which is expected to provide greater performance and reliability.
Expert Interview	Obtain expert opinion	Gather qualitative information regarding their technology and baseline and/or analogous systems.
Qualitative Assessment	A process to qualitatively evaluate your risk	A consistent method for qualitative evaluation of risk and the likelihood of risk occurrence. Usually this is done with expert opinion after some brainstorming. Risk is then documented. If quantitative assessment is to be added, this should be planned (see the next section).

Table 13.3
Guidelines for risk assessment

13.7 Guidelines for Risk Analysis (Step 3)

There are a number of mathematical methods in performing risk and decision analysis. Any reasonable analysis is better than no analysis. This is true for several reasons, mainly because an analysis brings more information to the decision process. Usually information leads to some sort of relative comparison or analysis. Absolute assessment can be avoided, and decisions can be based on historical baseline information.

One process of risk analysis is shown in Table 13.4. All of the steps in the table need to be performed.

The first goal in the analysis is to establish a parameter assessment (see Figure 13.7). The key parameters of concern are categorized in Table 13.4 with target and specification values. The present values are listed next. Expert opinion is sought, after which a mature estimate is made. In establishing an expert opinion, one must be realistic and understand whether the targets need to be reached with evolutionary or revolutionary technical advancement.

Figure 13.7
Performance parameter assessment

Subsystem: Transmitter				
Parameter	Target	Present Value	Mature Estimate	Prob. to Miss Target
Unit Cost	$225	$335	$240	10%
Power Rating	2 Watts	1.8 Watts	2 Watts	0%
Reliability	500,000 Hrs	200,000 Hrs	250,000 Hrs	60%
Schedule	1 Year	1.4 Years	11 Years	5%

Areas of Concern	Description	Guidelines
Risk Quantification	ID technology risks and categorize risks	This can be as simple as a ranking system or as complex as a full risk analysis. Mathematically, risk is the probability of Occurrence times the Severity of consequence (usually dollar value). Often requires analyzing expert opinion and quantifying data into probability distributions.
Performance	ID technical challenges associated with obtaining performance	Provide an assessment of risk associated with evolving a new design, which is expected to provide a greater level of performance and reliability. Establish target performance values, present values, and mature estimates.
Decision Analysis	Obtain expert opinion. Perform path analysis	This is the process of interviewing subject-area experts to gather qualitative information regarding their technology and baseline and/or analogous systems. Then a decision path should be established.
Jury Process	A process for quantifying each risk	This is a consistent method for qualitative evaluation of risk and the likelihood of risk occurrence. Expert-opinion jury process can rank probability of Occurrence and Severity cost to help quantify risk dollars.

Figures 13.8 and 13.9 provide guidelines for reviewing the area of risk that is actually involved. In the case of revolutionary technological advancement, it is most likely necessary to form a jury and fully judge the realities involved. Along with this mature estimate, the experts need to establish a probability of success or failure. Once this probability is established, a risk value can be obtained. This value is defined mathematically as the probability of failure times severity costs (see Figure 13.8).

In many cases, the cost of failure is the program value. In some cases, it can be higher, such as losing your customer or future programs. Once you are aware of your risk cost, you will be in an excellent position to start the risk-handling process and/or make decisions.

The most common analytic method for analyzing a decision is through decision path analysis. This is illustrated in Figure 13.10. The process is similar to evaluating risk. Each path has associated with it a failure probability and a cost. The total risk can be combined for each major path leg, and decisions can be based on the lowest risk path. Other factors may be difficult to work out, such as the risk of being too conservative. This can cause loss of future business as well. When evaluating a high-risk program, the best path is to try to advance your technology without losing your customer and a program's potential future. This can mean that there is often a need for risk handling at the highest management levels.

Table 13.4
Guidelines for risk analysis

Figure 13.8
Quantifying risk into dollars

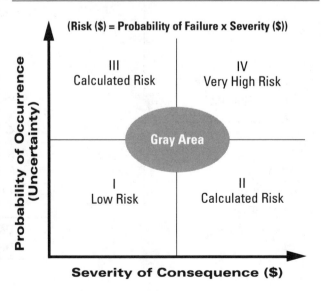

(Risk ($) = Probability of Failure x Severity ($))

Probability of Occurrence (Uncertainty)

III Calculated Risk

IV Very High Risk

Gray Area

I Low Risk

II Calculated Risk

Severity of Consequence ($)

Figure 13.9
*Quantifying
risk judgment*

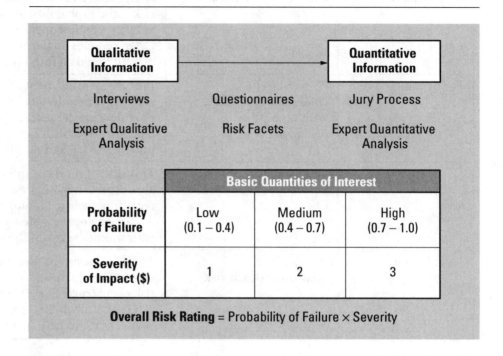

Figure 13.10
Decision path analysis

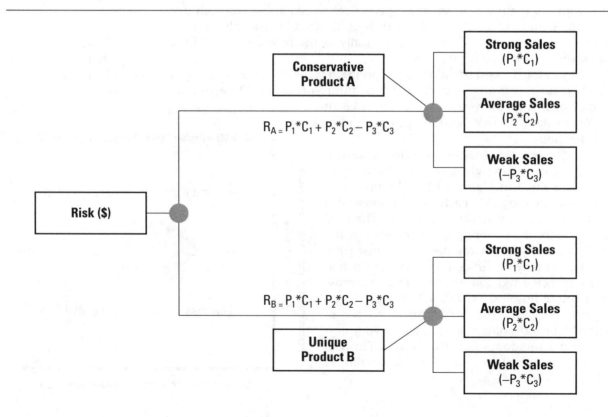

13.8 Guidelines for Risk Handling (Step 4)

With the information provided in the analysis, intelligent risk handling can be pursued. There are a number of ways to handle risk issues; each choice depends on the situation. Risk handling includes the major areas shown in compact notation of Table 13.5. These include risk avoidance, risk assumption, risk transfer, and risk control. In all cases, you should know your options. A watchlist should be developed that lists the program's risks, facets (areas of impact), and the handling actions.

This list may be expanded further for each item and the department that is handling the actions. At this point in your risk management, you should be in a reasonable position to manage potential problems without jeopardizing the impact they have on your customer.

Figure 13.11
Risk handling

- **Watchlist is an output of risk analysis areas of concern and risk priorities**
  - ✓ Make recommendations concerning risk avoidance, risk assumption, risk transfer, and risk control

- **Watchlist example:**

Event/Item	Area of Impact (Risk Facet)	Handling Action
Part A – low MTBF	Reliability	Use alternate part or implement corrective action; negotiate requirement
Loss of supplier	Production Cost	Seek second source
Long lead items delayed	Schedule	ID early in program; buy a place in line

Table 13.5
Guidelines for risk handling

Method	Description	Guidelines
Risk Avoidance	Avoiding unnecessary risks	Selecting the lowest risk choice using risk analysis.
Risk Assumption	Understanding and accepting known risks	Accepting risk at a specified safety level. For example, assume that the specification limit will be exceeded and negotiate with your customer.
Risk Transfer	Sharing risks	Sharing risk with contractors/customers through warranties, etc.
Knowledge & Research	Understanding technical risk issues	Understanding technical risk and reducing risk through skills and ingenuity.
Risk Control	Controlling risk through management	Continual monitoring and documenting progress on key milestones and corrective actions from the watchlists, enabling risk decisions to be optimally made in a timely manner.

CHAPTER 14

Thermodynamic Reliability Engineering

14.1 Thermodynamics and Reliability Engineering

Although reliability mathematics is well established, having probability theory as its basic tool, reliability science for physics-of-failure lacks a basic foundation. Thermodynamics is a natural candidate. Many engineers do not realize how closely tied thermodynamics is to reliability, since these subjects are treated separately. In this chapter, we apply the laws of thermodynamics and reliability theory to illustrate the key aspects that link these sciences into "Thermodynamic Reliability Engineering" (TRE) (see Reference 1) that helps in the understanding of reliability physics-of-failure problems.

When building a semiconductor component, manufacturing a steel beam, or simply blowing up a balloon, a system is created which interacts with its environment. Left to itself, the interaction between the system and environment degrades the system of interest. The degradation is driven by a tendency of the system to come to thermodynamic equilibrium with its environment. The total order of the system plus its environment tends to decrease. The air in the balloon will start to diffuse through the rubber wall; impurities from the environment will diffuse into otherwise more pure semiconductors; internal manufacturing stresses will cause dislocations to move into the semiconductor material; iron alloy steel beams will start to corrode as oxygen atoms from the atmospheric environment diffuse into the steel. In all of these cases, the spontaneous processes creating disorder are irreversible. For example, the air is not expected to go back into the balloon, the semiconductor will not spontaneously purify, and the steel beam will only build up more and more rust. The original order created in a manufactured product diminishes in a random manner and becomes measurable in our macroscopic world.

One finds that measurable disorder (aging) has occurred. In thermodynamics, the quantity entropy defines the property of matter that measures the degree of microscopic disorder that appears at the macroscopic level.

The *second law of thermodynamics* describes what is intuitively known about these systems in terms of *entropy*. That is, **the spontaneous process that takes place in the system-environment interaction when left to itself increases the total entropy.** The second law is another way of saying that the total order in the system plus the environment changes toward disorder.

Associated with the increase in total disorder, or entropy, is a loss of ability to do useful work. The total energy has not been lost but degraded. The total energy of the system plus the environment are **conserved** during the process when total thermodynamic equilibrium is approached. For the balloon example, prior to aging, the system energy was in a highly organized state. The energy could be released in the form of the kinetic energy of the balloon motion through the air. After aging, the energy of the gas molecules (which were inside the balloon) is now randomly distributed in the environment. These molecules cannot easily perform organized work; the steel beam, when corroded into rust, has lost its strength. These typical second-law examples describe the irreversible processes that cause aging. More precisely, if entropy has not increased, then the system has not aged.

In this sense, we define thermodynamic reliability engineering as **the act of recognizing, studying, and evaluating the potential for irreversible problems in a product, and the employment of this information into its design and/or the manner in which the design is used.**

14.2 The System and Its Environment

In thermodynamics, it is important to define both the system and its neighboring environment. The following definitions are used here for TRE:

- *The system is a portion of matter and/or a region of space set apart for study. From an engineering point of view, of concern is the possible aging of the system.*
- *The environment is the neighboring matter, which interacts with the system.*

It is not of interest to consider for the environment the totality of nature, but only that part which directly interacts with a given system. This interaction drives the system plus the environment toward a state of thermodynamic equilibrium.

14.2.1 Work and Free Energy

Prior to aging, the system has a certain portion of its energy that is "available" to do useful work. The available energy is called the *free energy*, ϕ. The system free energy is in practice less than the system energy, U; i.e., if T denotes the temperature of the environment and if S denotes the system entropy, then $\phi = U - TS$, which obeys $\phi < U$. If the system's initial free energy is denoted by ϕ_i (before aging) and the final free energy is denoted by ϕ_f (after aging), then $\phi_f < \phi_i$. The system is in thermal equilibrium with the environment when the free energy is minimized.

- *For an environment at a fixed temperature, the isothermal change in the system free energy is equal to the work done by the system on the environment,*

$$\text{Work} = (\phi_f - \phi_i) \qquad (14.1)$$

14.2.2 The Free Energy Roller-Coaster and the Arrhenius Law

Sometimes, the system path to the free-energy minimum is smooth and downhill all the way to the bottom. For other systems, the path may descend to a relative minimum, but not an absolute minimum, somewhat resembling a roller-coaster. The path goes downhill to what looks like the bottom and faces a small uphill region. If that small hill could be scaled, then the final drop to the true minimum would be just over the top of the small hill. The small climb before the final descent to the true minimum is called a free energy barrier. The system may stay for a long period of time in the relative minimum before the final decay to true equilibrium.

Often the time spent in the neighborhood of the relative minimum is the lifetime of a fabricated product, and the final descent to the true free energy minimum represents the catastrophic failure of the product. The estimated lifetime, τ, in which the system stays at the relative minimum obeys the Arrhenius law $(1/\tau) = 1/\tau_o exp (-\Delta/K_B T)$, where Δ is the height of the free energy barrier.

14.2.3 Thermodynamic Work and the First Law

As a system ages, work is performed by the system on the environment or vice versa. Measuring the work isothermally performed by the system on the environment, and if the effect on the system could be quantified, then a measure of the change in the system's free energy could be obtained.

The bending of a paper clip back and forth illustrates cyclic work done by the environment on the system that often causes dislocations to form in the material. The dislocations cause metal fatigue and, thereby, the eventual fracture in the paper clip; the diffusion of contaminants from the environment into the system may represent chemical work done by the environment on the system. We quantify such changes using the first law of thermodynamics. The first law is a statement that energy is *conserved* if one regards heat as a form of energy. The *first law of thermodynamics* describes conservation of energy.

- *The energy change of the system, ΔU, is partly due to the work, ΔW, performed on the system by the environment, and partly due to the heat, ΔQ, which flows from the environment to the system,*

$$\Delta U = \Delta Q + \Delta W \qquad (14.2)$$

If heat flows from the system to the environment, then our sign convention is that $\Delta Q < 0$. Similarly, if the work is done by the system on the environment, then our sign convention is that $\Delta W < 0$.

Applied to the TRE example, heat is released into the environment when a paper clip is bent back and forth. The work done on the paper clip results in a plastic deformation. In this case, the portion of the work that caused plastic deformation is $\Delta W_A = \Delta U_{plastic}$ (A indicating Aging). After many bends back and forth, the plastic energy (as well as the entropy) builds up until the system goes into catastrophic failure. In the free-energy description, the bending back and forth sends the system over the free energy barrier.

14.2.4 The Free Energy and the Second Law

We have defined a thermodynamic quantity called the free energy as that quantity of energy which is available to the system to perform isothermal work on the environment. The *second law* can also be described using the free energy for TRE:

- The spontaneous process that takes place over time in a system immersed in an isothermal (constant temperature) environment decreases the free energy of the system toward a minimum value. The spontaneous process reduces the ability of the system to perform useful work on the environment, which results in system aging. Mathematically, this situation is shown in Figure 14.2 and is discussed in the next section.

The first and second laws have now been defined using thermodynamic reliability engineering examples. With these definitions, we are in a position to proceed with a discussion of the aging process.

14.3 The Aging Process

The irreversible mechanisms of interest here that cause aging are activation, diffusion, and external force-induced process (see Reference 2) as shown in Figure 14.1. Combinations of these processes provide complex forms of aging. Aging depends on the rate-controlling process. Any one of these three processes may dominate depending on the failure mode. Alternately, the aging rate of each process may be on the same time scale, making all such mechanisms equally important.

The notions of reversible and irreversible processes define two regimes called equilibrium and nonequilibrium thermodynamics. Equilibrium thermodynamics provides methods for describing the initial and final equilibrium system states without describing the details of how the system evolves to a final equilibrium state. Such final states are those of maximum total entropy (for the system plus environment) or minimum free energy (for the system).

Nonequilibrium thermodynamics describes in more detail what happens during the evolution to the final equilibrium state, e.g., the precise rate of entropy increase or free energy decrease. Those parts of the energy exchange

Figure 14.1

Three main types of aging processes

Aging can be due to:

✓ Forced process

✓ Activation

✓ Diffusion

Figure 14.2
Thermodynamic states

broken up into heat and work by the first law are also tracked during the evolution to an equilibrium final state. At the point where the irreversible processes virtually slow to a halt, the process approaches reversibility. Mathematically, this is described in Figure 14.2.

For example, as the work is performed by a chemical cell (a battery with an electromotive force), the cell ages and the free energy decreases. Nonequilibrium thermodynamics describes the evolution which takes place as current passes through the battery, and the final equilibrium state is achieved when the current stops and the battery is dead. "Recharging" can revive a secondary battery.

14.3.1 State Variables

Thermodynamics also provides a natural way to define a system's state through macroscopic state variables such as temperature, volume, and pressure. These macroscopic parameters depend on the particular system under study and can include voltage, current, electric field, vibration displacements, and so forth. These are all thermodynamic state variables. Thermodynamic parameters can be categorized as intensive or extensive. Intensive variables have uniform values throughout the system such as pressure or temperature. Extensive variables are additive such as volume or mass. For example, if the system is sectioned into two subsystems, the total volume, V, is equal to the sum of the volumes of the two subsystems. The pressure is intensive. The intensive pressures of the subsystems are equal and the same as before the division. Intensive parameters can be defined in the small neighborhood of a point.

Some pairs of state variables are directly related to mechanical work. Examples of mechanical work variables are provided in the table below.

Table 14.1
Generalized thermodynamic state variables for mechanical work

Common Systems	Generalized Force Y	Generalized Displacement X	Mechanical Work dW = YdX
Gas	Pressure (−P)	Volume (V)	−P dV
Chemical Potential	Chemical potential (μ)	Molar number of atoms or molecules (N)	μ dN
Spring	Force (f)	Distance (x)	f dx
Mechanical Wire/Bar	Tension (J)	Length (L)	J dL
Mechanical Strain	Stress (S)	Strain (e)	S de
Electric Polarization	−π	E	−π dE
Capacitance	Voltage (V)	Charge (q)	V dq
Induction	Current (I)	Magnetic flux (Φ)	IdΦ
Magnetic Polarizability	Magnetic intensity (H)	Magnetization (M)	H dM
Linear System	Velocity (v)	Momentum (m)	v dm
Rotating Fluids	Angular velocity (ω)	Angular momentum (L)	ωdL

14.3.2 Test Design by Failure Modes and Aging Stresses

In reliability testing, it is important to know which thermodynamic quantities will accelerate potential failure modes. Table 14.2 provides a very generalized overview. This table, while assembled with careful research into these processes, is by no means a definitive characterization. The table can be used both as a guide for categorizing stresses related to failure modes/mechanisms and as a format. As a format, a product can initially be analyzed using this tabular method to analyze potential failure modes, and then accelerated tests can be designed using relevant stresses as shown in Example 14.1. This is termed here Test Design by Failure Modes. Figure 14.3 also illustrates some mechanisms and related aging rates. The figure illustrates that aging rates are both parametric and catastrophic, and some general areas where the three mechanisms – diffusion, activation, and forced – apply. Examples are provided of each in this chapter.

Table 14.2

Some common thermodynamic aging mechanisms and related thermodynamic variables

Aging Mechanism/ Failure Modes	Thermodynamic Variable/Acceleration Stress Factor									
	Type of Aging	Test	Temp	Humidity	Electric Potential	Temp Cycle	Current Density	Static, Dynamic, or Vibration Pressure	Shock Pressure	Corrosion Chemical Potential
Oxidation	D, A	HTL	√							√
Chemical Reactions	A	THB	√	√						√
Fatigue	F	TC, Vib, Shk	√			√		√	√	
Corrosion	A, D, F	THB	√	√	√					√
Fracture	F	Vib, Shk	√	√				√	√	
Wear	F	Mech Cycle	√	√				√	√	
Corrosion Fatigue	A, D, F	THB, Vib	√	√	√	√		√	√	√
Stress Corrosion Cracking	A, D, F	THB, Vib	√	√	√	√		√	√	√
Electromigration	A, D, F	Current Density	√				√			
ESD	F	ESD	√		√					
Dielectric Breakdown	A, F	HV	√		√					

Table Key: Column 2: F = Forced, A = Activated, D = Diffusion

Column 3: HTOL = High-Temperature Operating Life, THB = Temperature-Humidity-Bias, TC = Temperature Cycle, Vib = Vibration, Shk = Shock, HV = High Voltage

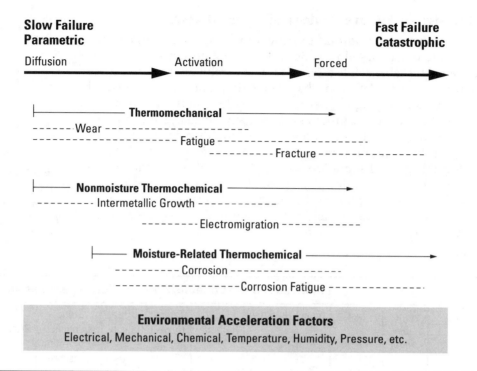

Slow Failure
Parametric

Fast Failure
Catastrophic

Diffusion → Activation → Forced →

Thermomechanical →
------ Wear ----------------------
-------------------- Fatigue --------------------------
-------------- Fracture ----------------

Nonmoisture Thermochemical →
--------- Intermetallic Growth -----------
--------- Electromigration -----------

Moisture-Related Thermochemical →
----------- Corrosion ------------------
------------------ Corrosion Fatigue --------------

Environmental Acceleration Factors
Electrical, Mechanical, Chemical, Temperature, Humidity, Pressure, etc.

Figure 14.3
Aging rates and some related mechanisms

Table 14.3
Microswitch example of accelerated stresses

▼ **Example 14.1** *Test design by failure modes for a microswitch*

Problem:
A simple microswitch is to be used in a telephone. Using the test design by failure-mode method, determine the appropriate accelerated stress tests.

Solution:
Ideally, a brainstorming session should be held to analyze potential failure modes and mechanisms for the microswitch. Table 14.3 provides an example of possible tabulated failure modes and mechanisms versus related stresses. In the table, one can also include a ranking factor as shown to assess engineering judgment of the most likely problems.

Aging Mechanism/ Failure Modes	Thermodynamic Variable/Acceleration Stress Factor								
	Rank Factor	Test	Temp	Humidity	Electric Potential	Temp Cycle	Static, Dynamic, or Vibration Pressure	Shock Pressure	Corrosion Chemical Potential
Oxidation (intermittent contact)	3	HT	√						√
Fatigue (mechanical break)	1	Mech Cycle	√			√	√	√	
Corrosion & Corrosion Fatigue (intermittent contact)	4	THB	√	√	√				√
Wear (mechanical break)	2	Mech Cycle	√	√			√	√	

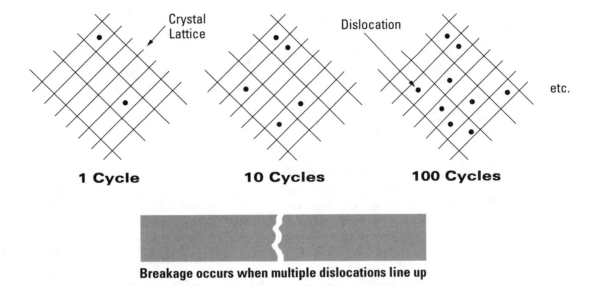

1 Cycle **10 Cycles** **100 Cycles** etc.

Breakage occurs when multiple dislocations line up

Fatigue and wear are ranked as the most likely failure mechanisms to occur. All potential failure mechanisms should be tested. However, most of the test allocation should be concentrated on mechanical cycling of the switch. Therefore, tests should include a high-temperature bake, temperature-humidity-bias, and mechanical cycling of the switch. Since the failure mechanisms can be interrelated, the same switches should be used on all tests.

Figure 14.4
Dislocations in crystal lattice of a metal during fatigue

14.4 Aging Due to Cyclic Force

Thermodynamics is commonly used to interpret cyclic forced processes in which heat and work are interchanged. Simple reliability examples can be related to aging using this thermodynamic framework. Earlier, a simple example of cyclic work performed by the bending of a paper clip back and forth was described. The cyclic thermodynamic work is converted into heat, which goes in part to increase the entropy of the system as dislocations are added every cycle. Work is also converted into heat released into the environment. Cyclic work done on the paper clip in part results in plastic deformation in the form of dislocations created in the material. These produce metal fatigue. This is illustrated in Figure 14.4. Aging from such fatigue is due to external forces, which eventually result in fracture of the paper clip.

Cyclic work can be illustrated graphically as shown in Figure 14.5 where the generalized displacement, X, and force, Y, have been employed. The work "areas" in Figure 14.5 describe the work done per cycle on the system. For example, for the paper clip, the coordinates are stress and strain (the forces constitute the strain). Examples of other thermodynamic mechanical work variables are also shown in Table 14.1.

Cyclic work is repetitive and can be summed as a measure of the total cyclic work performed. In this chapter, we define cumulative damage by the ratio of the sum of the thermodynamic work performed per cycle to the total amount of cyclic work required to cause failure as (see References 1 and 2),

$$\text{Damage} = \frac{\text{Thermodynamic Work for n Cycles}}{\text{Thermodynamic Work to Failure}} \qquad (14.3)$$

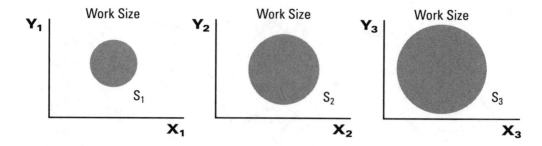

Figure 14.5

Cyclic work plane of three different sizes

14.4.1 A Derivation of Miner's Rule

The above expression for damage may be used to derive Miner's rule (see Reference 3) commonly used for accumulated fatigue damage and a number of other useful expressions in reliability engineering. It would perhaps be more accurate to write the damage in terms of the number of dislocations produced during each cycle. However, this number is usually unknown. Therefore, we will proceed with the above equation. We start with the thermodynamic work, W, that is a function of the cyclic size, S, and the number of cycles, n, such that

$$W_n = W(S, n) \qquad (14.4)$$

If we assume that the work for n cycles of the same size obeys the relationship,

$$W_n \cong n\,W(S) \qquad (14.5)$$

failure will occur suddenly after n cycles. Substituting this into the equation for damage, and summing over cycles of possibly different size yield,

$$Damage = \frac{n_1 W(S_1) + n_2 W(S_2) + \ldots}{W_{Failure}} = \frac{n_1 W(S_1)}{W_{Failure}} + \frac{n_2 W(S_2)}{W_{Failure}} + \ldots \qquad (14.6)$$

If N_j denotes the number of cycles of size S_j to failure, then the identity,

$$W_{Failure} = N_1 W_1(S_1) = N_2 W_2(S_2) = N_3 W_3(S_3) = \ldots \qquad (14.7)$$

allows us to obtain the classical fatigue equation due to Miner for cumulative damage at each ith stress level that reads,

$$Damage = \frac{n_1 \, W(S_1)}{N_1 \, W(S_1)} + \frac{n_2 \, W(S_2)}{N_2 \, W(S_2)} + \frac{n_3 \, W(S_3)}{N_3 \, W(S_3)} + \ldots = \frac{n_1}{N_1} + \frac{n_2}{N_2} + \frac{n_3}{N_3} + \ldots = \sum_i \frac{n_i}{N_i} \qquad (14.8)$$

This derivation (see Reference 2), which is not in Miner's original paper (see Reference 3), helps in understanding the use of $W_n \cong n\,W(S)$ in some high-stress applications.

▼ **Example 14.2** *Miner's rule*

Problem:

A pressure vessel is made of aluminum alloy and operates in two states at 20 cycles per minute. It operates in the first state 60% of the time at a stress change of 1500 psi; the cycles to failure in this state are 1,000,000 cycles. In the second state (40% of the time), a stress change of 2400 psi is exerted on the vessel, and the cycles to failure in this state are 215,000. The problem is to find the expected life in hours of this hydraulic pressure unit.

Solution:

The information is $n_1 + n_2 = 20$ cycles per minute, or effectively $n_1 = 0.6 \times 20 = 12$ cycles per minute and $n_2 = 0.4 \times 20 = 8$ cycles per minute.

From Miner's rule, we have

$$Damage = \sum_i \frac{n_i}{N_i} = \frac{n_1}{N_1} + \frac{n_2}{N_2} \qquad (14.9)$$

which is the portion of total life consumed per minute of operation. The damage per unit time is

$$Damage = \frac{12}{1,000,000} + \frac{8}{215,000} = 0.49 x 10^{-4} \text{ per minute} = 0.00294 \text{ per hour}$$

Since failure occurs when damage = 1, then $Damage = 1 = $ (at N hours for failure) $\times$ (0.00294 amount of damage per hour) or solving for N hours is N hours for failure = 1/0.00294 = 340 hours.

This example may be found in Miner's original paper in Reference 3.

▼ Example 14.3

Problem:
Aluminum alloy has the following fatigue characteristics:
 Stress 1, N = 45 cycles,
 Stress 2, N = 310 cycles, and
 Stress 3, N = 12,400 cycles.

How many times can the following sequence be repeated?
 n_1 = 5 cycles at stress 1,
 n_2 = 60 cycles at stress 2, and
 n_3 = 495 cycles at stress 3.

Solution:
The fractions of life exhausted in each block are

$$\frac{n_1}{N_1} = \frac{5}{45} = 0.111, \quad \frac{n_2}{N_2} = 0.194, \quad \frac{n_3}{N_3} = 0.04$$

The fraction of life exhausted in a complete sequence is approximately 0.345. The life is entirely exhausted when $Damage = 1$, and at this point the sequence is repeated X times; X(0.345) = 1. Solving for X gives X = 2.9 times. For example, after the sequence is repeated twice, the fraction of life exhausted is 0.69. Then n_1 and n_2 would bring it up to about 0.995, which is close enough for failure to occur.

Assumption in Miner's Rule

 Miner's rule applies if cyclic work can be approximated by the number of cycles times an average amount of work per cycle of a certain size [i.e., $W_n \cong nW(S)$]. However, if the work is some nonlinear function of the number of cycles, then Miner's rule is invalid. If the paper clip bends past its elastic limit in some of the stress cycles, then W_n is nonlinear in the material. This corresponds to overstressing the material. If the material is severely overstressed during each cycle, then the approximation $W_n \cong nW(S)$ used in the derivation of Miner's rule is not reasonable. In this case, the amount of work to bend the paper clip at the same stress level will be largely reduced from cycle to cycle and W_n is nonlinear in n. Overall, the assumptions in Figure 14.6 follow.

Figure 14.6
Assumption in Miner's rule

Miner's Rule Assumes:
 Work = W(S_i) n

Sometimes Require:
 Work = W(S_i, n)

In the latter case, the work could be incrementally summed with the actual stress applied to the material. Damage is then expressed in terms of the generalized coordinates in Figure 14.5 as

$$Damage = \frac{\sum_n \oint Y_n \, dX_n}{W_{Failure}} \qquad (14.10)$$

Thermodynamic Extensions of Miner's Rule

Once the thermodynamic work argument is appreciated, then it becomes evident that Miner's rule will be of use in the reliability engineering to any system capable of undergoing cyclic thermodynamic work (see Reference 14). For example, in addition to metal fatigue, Miner's rule should follow in the case of chemical cells, e.g., secondary batteries. It is known that battery capacity life is extended as the depth of discharge (e.g., stress level) decreases (see Reference 14). In the chemical cell case, the cyclic "area" is the voltage-charge plane (internal disorder parameter is the degree of corrosion instead of dislocations as in metal fatigue). From Table 14.1, one might imagine other extensions such as magnetic hysteric behavior in magnetic recording media. Here, the external cyclic work is the area in the magnetization-magnetic intensity plane.

14.4.2 A Derivation of the Fatigue Time Compression Factor Used in Temperature Cycle Derivation

In temperature cycling, a temperature change, ΔT, in the environment, from one extreme to another, causes expansion and/or contraction (i.e., strain) in a material system. The strain, e, is accompanied by stress, S, in the material. The thermodynamic work in a cycle is

$$\Delta W = \oint S de \qquad (14.11)$$

Assuming that the nonlinear part of the strain can be approximated by a power function $S = R(e_p)^k$, where R is a material constant, p is a subscript that indicates plastic strain, e (rather than elastic), and k is the power-law exponent, and assuming plastic strain on only one part of the cycle, then

$$\Delta W = \frac{R}{k+1}(\Delta e_p)^{k+1} \qquad (14.12)$$

where the integral has been taken over the plastic strain. The plastic strain, Δe_p, is a function of ΔT. Furthermore, assume that $\Delta e_p = (B \Delta T)^v$, where v is an exponent (often $v = 1$) and B is a constant. The total work for n_1 cycles with temperature variation, ΔT_1, is given by

$$\Delta W_1 = \frac{n_1 R}{k+1}(B\Delta T_1)^K \qquad (14.13)$$

where $K = v(k+1)$. The ratio of the cyclic fatigue work for a given material between two different temperature environments 1 and 2 is

$$\frac{\Delta W_1}{\Delta W_2} = \frac{n_1}{n_2}\frac{(\Delta T_1)^K}{(\Delta T_2)^K} \qquad (14.14)$$

Let n_1 be the number cycles to failure in environment 1 and similarly for n_2 in environment 2. At the point of failure, the total work will be the same for either environment $(\Delta W_1)_F = (\Delta W_2)_F$, giving

$$\frac{N_2}{N_1} = \left(\frac{\Delta T_1}{\Delta T_2}\right)^K \qquad (14.15)$$

This ratio is the temperature cycle "Coffin-Manson" acceleration factors in Chapter 9, Figure 9.4. This empirically based power-law expression is thereby shown derived here from the TRE damage expression and the plastic thermally induced strains described above.

14.4.3 A Mechanical Cycle (Vibration) Fatigue Time Compression Derivation

In a similar manner to the above derivation, we can exhibit the mechanical cyclic (vibration) acceleration factor. The above method can be used with cyclic stress due to vibration. In this case, we approximate $\Delta e_p = (\beta\, G)^\gamma$ where γ is an exponent, β is a constant, and G is the vibration G_{rms} input strength. The work reads

$$\Delta W_1 = \frac{n_1 R}{k+1}(\beta\, G)^b \qquad (14.16)$$

where $b = \gamma(k + 1)$. Again, the ratio of the cyclic fatigue work is considered, now between two vibrational G environments 1 and 2 performed on the same material. This is

$$\frac{\Delta W_1}{\Delta W_2} = \frac{n_1}{n_2}\frac{(G_1)^b}{(G_2)^b} \qquad (14.17)$$

At the point of failure, the amount of work is the same for either environment $(\Delta W_1)_F = (\Delta W_2)_F$ requiring $n_1 = N_1$ cycles to failure in environment 1 and $n_2 = N_2$ cycles to failure in environment 2. The ratio is

$$\frac{N_2}{N_1} = \left(\frac{G_1}{G_2}\right)^b \qquad (14.18)$$

This is a commonly used relation for cyclic compression. Since the number of cycles is related to cycle frequency, f, and the period, T, according to

$$N = fT \qquad (14.19)$$

if f is constant, then

$$\frac{T_{n2}}{T_{n1}} = \frac{N_2}{N_1} \qquad (14.20)$$

from which follows

$$\frac{T_{n2}}{T_{n1}} = \left(\frac{G_1}{G_2}\right)^b \qquad (14.21)$$

This can be related to random vibration through the power spectral density W_{psd}. For example, when G_{rms} can be approximated as (for a flat spectra with bandwidth Δf)

$$G_{rms} \cong \sqrt{W_{psd}\,\Delta f} \qquad (14.22)$$

then the time compression expression is

$$\frac{T_2}{T_1} = \left(\frac{W_{psd1}}{W_{psd2}}\right)^{b/2} = \frac{N_2}{N_1} \qquad (14.23)$$

This is a common form used for the random vibration acceleration factor. This is, of course, the vibration acceleration factor in Figure 9.6 of Chapter 9.

▼ Example 14.4 *Relation to the S-N curve*

Problem:

The experimental number of cycles to failure, N, for a given stress, S, level constitutes S-N curve data. Such data are widely available in the literature for many common materials. The slope of the S-N curve provides an estimate of the exponent b above. This also determines the exponent in the time-compression expression above between vibration environments. Show that the W_{psd} time compression exponent is related to the S-N curve exponent divided by 2 (as noted in Chapter 9; also see Example 9.5).

Solution:

We write the cyclic equation for N_2 as where $G \propto S$:

$$N_2 = \left(\frac{G_1}{G_2}\right)^b N_1 = C_1 G_2^{-b} = C\ S^{-b}$$

For illustrative purposes, only one environment is considered. Therefore, we have equated $N_1(G_1)b$ to a constant. The relationship is generally used to analyze S-N data.

$$N = C\ S^{-b}$$

Therefore, the S-N exponent b is the W_{psd} time-compression exponent divided by 2 (i.e., $S \propto G \propto W^{0.5}$).

Figure 14.7
A simple corrosion cell

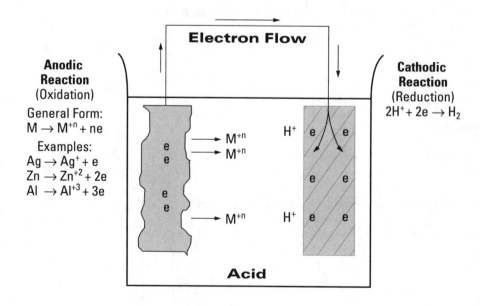

Anodic Reaction (Oxidation)
General Form:
$M \rightarrow M^{+n} + ne$
Examples:
$Ag \rightarrow Ag^+ + e$
$Zn \rightarrow Zn^{+2} + 2e$
$Al \rightarrow Al^{+3} + 3e$

Electron Flow

M^{+n}
M^{+n}
M^{+n}

H^+

Cathodic Reaction (Reduction)
$2H^+ + 2e \rightarrow H_2$

Acid

14.5 Corrosion and Activation

In this section, the corrosion process is considered to illustrate the basic principles of complex aging. Common corrosion in a chemical battery requires the presence of four elements: a metal anode, a cathode, an electrolyte, and a conductive path as shown in Figure 14.7.

Figure 14.8 illustrates that a similar electrochemical process can be formed with an electrolyte on a simple metal surface. Small irregularities on the surface can form cathode, C, and anode, A, areas, usually due to differences in oxygen concentration. The exchange of matter can be described in nonequilibrium thermodynamics in terms of the currents at the electrodes. In the general theory, the forward current, I_f, leaves the anode electrode into the electrolyte, and the backward current, I_b, enters the cathode (see Reference 4). These currents are given by the products of the rate constants and relevant concentrations at the electrode surface as

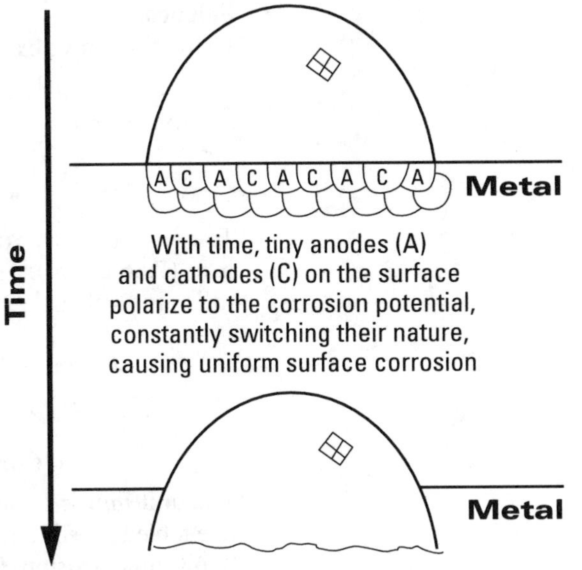

$$I_f = nHAK_fC_o \text{ and } I_b = nHAK_bC_R \qquad (14.24)$$

where n is the number of electrons involved in the reaction, A is the electrode area, H is Faraday's constant (96,500 Coulombs/eq.), K_f and K_b are temperature-dependent rate constants discussed below, and C_O and C_R are the concentrations at the electrode surface (not necessarily equal to the bulk concentrations). The total net corrosion current is

$$I = I_f + I_b \qquad (14.25)$$

The mass, dM, transferred in time, dt, in the reaction can be thought of as an aging parameter for the material. According to Faraday's Law, (dM/dt) is proportional to the net current as

$$dM \text{ (mass of metal dissolving (g))} = kIdt \qquad (14.26)$$

where

$$k = \left(\frac{A_m}{nH}\right) = \frac{\text{atomic wt of metal}}{\text{no. of electrons transferred} \times 96,500 \text{ A-sec}} \qquad (14.27)$$

Thus, A_m is the atomic mass. The corrosion rate (dM/dt) is

$$Corrosion \ rate = \left(\frac{A_m}{nH}\right)I \qquad (14.28)$$

and is proportional to the current flowing.

Figure 14.8
Uniform electrochemical corrosion depicted on the surface of a metal

With time, tiny anodes (A) and cathodes (C) on the surface polarize to the corrosion potential, constantly switching their nature, causing uniform surface corrosion

▼ **Example 14.5** *Corrosion rate*

Problem:
Consider iron corroding in air-free acid at an electrochemical corrosion rate of $1\mu A/cm^2$. It dissolves as ferrous ions (Fe^{+2}) and thus, $n = 2$. Obtain the corrosion rate in mils per year.

Solution:

Using the above expression

$$Corrosion\ rate = \frac{55.8 g/equivalent\,(10^{-6} coul/sec-cm^2)}{2(96,487\,\frac{couls}{equivalent})} = 2.89x10^{-10}\,\frac{g}{sec-cm^2}$$

To convert the corrosion rate to mils per year, first divide by the density of iron ($7.86\ g/cm^3$). Additionally there are 3.15×10^7 seconds/year and 393.7 mils/cm (= 1 inch/2.54 cm × 1000 mils/inch), then

$$Corrosion\ rate = 2.89x10^{-10}\,\frac{g}{sec-cm^2}\,x\,\frac{1}{7.86\frac{g}{cm^3}}\,x\,3.15x10^7\,\frac{sec}{year}\,x393.7\,\frac{mils}{cm}$$

$$Corrosion\ rate = 0.456\ mils\ per\ year\ (mpy)$$

The unit *mpy* is a common corrosion rate unit. Table 14.4 below provides a relative table for estimating corrosion resistance of materials found in a number of books on corrosion. From the table, note that 0.46 mils per year is an outstanding corrosion rate, indicating an excellent corrosive-resistive material. Note this is a relative table established under a certain set of environmental conditions.

▼ **Example 14.6** *Corrosion rate equation*

Problem:

Provide a simple metric equation expression for the corrosion rate, *R*, in cm/hr with parameter of corrosion time, *t*, in hours, weight, *W*, in grams, density, *D*, in grams/cm³, and area, *A*, in cm². Then convert this expression to a more commonly used mixed-units corrosion expression where *R* is in mils/yr, *A* is in inch², *t* is in hrs, *D* is in g/cm³, and *W* is in mg.

Solution:

When the corrosion rate is linear, the rate can be found by simply dividing the mass corroded by the corrosion time. In units of *R(cm/hr)*, this is the mass, *W*, with exposed area, *A*, in a laboratory experiment as

$$R(cm/hr) = \frac{W(grams)}{t(hrs)D(g/cm^3)A(cm^2)}\qquad(14.29)$$

Table 14.4
Relative corrosion resistance

Here, the expression can be checked using its units. Now converting to a mixed-unit expression, the conversion factor $R(cm/hr)(3.44 \times 10^6$ mils/cm × hr/yr) gives *R* in mils/yr. Thus, $R(cm/yr) = 1/(3.44 \times 10^6)\ R(mils/yr)$. Since

Relative Corrosion Resistance	Mils Per Year (mpy)	Micrometer Per Year mm Per Year
Outstanding	<1	<25
Excellent	1 – 5	25 – 100
Good	5 – 20	100 – 500
Fair	20 – 50	500 – 1000
Poor	50 – 200	1000 – 5000
Unacceptable	200+	5000+

$A(cm^2)(inch^2/(2.54\ cm)^2)$ gives A in $inch^2$, then $A(cm^2) = 6.45\ A(inch^2)$. Similarly, $W(mg)/1000 = W(mg)$. Inserting these values above,

$$R(mils/yr) = \frac{3.44x10^6\ \dfrac{W(mg)}{1000}}{t(hrs.)D(g/cm^3)6.45A(inch^2)} = \frac{534\ W(mg)}{t(hrs.)D(g/cm^3)A(inch^2)}$$

This is one common mixed-unit expression used in corrosion engineering.

14.5.1 The Corrosion Rate Constants

The free energy change accompanying an electrochemical reaction is usually expressed by the change in the Gibbs free energy, G, which is commonly used as the thermodynamic potential for chemical reaction. For a reaction, it can be calculated by the following equation

$$\Delta G = -nHE \qquad (14.30)$$

where

ΔG = Gibb's free energy change,
n = number of electrons involved in the reaction,
H = Faraday constant (described above), and
E = cell potential.

Then the reaction rate is related to the free energy as

$$K = K_o exp\left\{-\frac{\Delta G}{RT}\right\} \qquad (14.31)$$

Thus, the magnitude of the rate constant depends on the electrode potential. This dependence is usually described by assuming that a fraction of the electrode potential, αE, is involved in driving the reduction process, while the fraction $(1-\alpha)E$ is effective in making the reoxidation process more difficult. Rate constants K_f and K_b (see References 2 and 4) are given by

$$K_f = K_{fo}exp\left\{-\frac{\alpha nHE}{RT}\right\} \text{ and } K_b = K_{bo}\,exp\left\{-\frac{(1-\alpha)nHE}{RT}\right\} \qquad (14.32)$$

where R = 8.314 J/mole K is the gas constant that is related to Boltzmann's constant K_B, where $R/N_o = K_B$ and N_o is Avogadro's number. Thus, ΔG represents the thermodynamic free energy for the reaction. The magnitude of energy represents a barrier height that must be surmounted to cause a corrosive transition state. Under equilibrium conditions, no net current flows and

$$I_f = I_b \qquad (14.33)$$

Inserting the forward and backward currents as originally defined

$$C_oK_{fo}\,exp\left\{-\frac{\alpha nHE}{RT}\right\} = C_RK_{bo}\,exp\left\{-\frac{(1-\alpha)nHE}{RT}\right\} \qquad (14.34)$$

Rearranging terms, the important Nernst equation can be obtained in terms of the concentrations (also can be written in terms of the chemical activities) governing the thermodynamics of the electrochemical reaction in terms of the concentrations (see Reference 5)

$$E = E_o + \frac{RT}{nF}\ \ln\left(\frac{C_o}{C_R}\right) \text{ or } G = G_o - RT\ \ln\left(\frac{C_o}{C_R}\right) \qquad (14.35)$$

Figure 14.9

Activation and corrosion thermodynamics

Nonequilibrium Thermodynamics (Aging Rate)

$$\frac{dMatter}{dt} \propto I_{O,R} \propto K_O Exp\left\{\frac{\Delta G}{RT}\right\}$$

Corrosion Current

Arrhenius Term

Equilibrium Thermodynamics ($I_0 = I_R$)

$$\Delta G = \Delta G_O + RT \ln\left(\frac{C_O}{C_R}\right)$$

Corrosion Tendency
• Corrode $\Delta G < 0$
• Will not corrode $\Delta G > 0$

where $E_o = RT/nF \ln\{K_{fo}/K_{bo}\} = -G_o/nF$ and is the reaction standard potential. The Nernst equation enables the calculation of the thermodynamic electrode potential when concentrations are known. It also can indicate the corrosive tendency of the reaction. When the thermodynamic free energy of the process is negative, there is a spontaneous tendency to corrode.

Corrosion is an interesting process because it demonstrates important thermodynamic facets of aging. Corrosion shows the importance of the activation process via the Arrhenius relationship, the electropotential external stress, and concentration. Concentration is not only important in electrochemical corrosive aging but also in diffusion. The above equations have illustrated aging due to reaction at the electrode interface.

14.5.2 A Derivation of Peck's Humidity Model

In this section, we provide a derivation for Peck's equation (see Chapter 9, Section 9.6). Peck's expression is an acceleration model for microelectronic corrosion. In microelectronics, surface corrosion is a function of the local relative humidity. The rate of corrosion and the rate of mass transport are related to the local relative humidity present at the surface. The reaction rate depends on concentration, C, according to the **chemical differential rate law** (also, law of mass action). For example, in reaction

$$3A + 2B \rightarrow D + E \tag{14.36}$$

the differential rate law may have concentrations [A], [B], and [D]. The rate

$$K(C) = \frac{d[D]}{dt} = k[A]^n[B]^m \tag{14.37}$$

where n or m are power exponents with orders of A and B, respectively. The brackets, such as those around *[A]*, indicate the concentration in moles per liter of A; $K(C)$ is then the reaction rate as a function of concentration. The overall order of the reaction (n and m) cannot be predicted from the reaction equation but must be found experimentally.

In terms of microelectronics, surfaces have an affinity for local relative humidity near the surface. This feeds the thin-film electrolyte, which affects the reaction rate both in terms of concentration in the anodic and cathodic reactions and also in terms of the rate of mass transport. For many corroding metals, the cathodic reduction of water itself in the electrolyte ($2H_2O \rightarrow OH^- + H_2 - e^-$) is a rate-controlling process (see Reference 15). The overall chemical reaction rate, if it could be described in simple terms as provided above, is thus some function of the local relative humidity. We will therefore use a somewhat naive approach by assuming that the relative humidity, similar to the formulation for $K(C)$ in the chemical differential rate law, has an overall rate that goes as a power functional form with the local percent Relative Humidity *(RH)* itself as

$$K\{C\} \propto (rh)^m \tag{14.38}$$

where $rh = RH/100$. We will see that this assumption is consistent with corrosion kinetics of Peck's expression where Peck's expression is applicable.

[Applicability may be as high as > 60%RH, depending upon the corrosion occurring. Some metals, such as iron, do not corrode below a certain relative humidity value (see Reference 15).] To show that this expression is consistent with Peck's expression, we first insert this assumption for $K(C)$ into our expressions for the corrosion currents as

$$I_f = (rh)^m\ nHAK_fC_o \quad \text{and} \quad I_b = (rh)^m\ nHAK_bC_R \tag{14.39}$$

(note that when $rh = 1$, the original expression results). The net current $I = I_f + I_b$ is zero under equilibrium conditions. For situations where the net current is not zero, the net current approaches that of either the forward or backward current, depending on the dominating mechanism. For example, anodic corrosion usually dominates a corrosion process. In this case, I is approximately I_f and the corrosion current is

$$I = (rh)^m\ n\,H\,A\,C_o\,K_{fo}\,exp\left\{-\frac{\Delta G_f}{RT}\right\} \tag{14.40}$$

In accelerated testing, the acceleration factor between a stress and use environment having different temperature and humidity conditions can be found from the rate ratio

$$AF_{TH} = \frac{Corrosion\ rate_{Stress}}{Corrosion\ rate_{Use}} = \frac{\left(\frac{A_m}{nH}\right)I_{Stress}}{\left(\frac{A_m}{nH}\right)I_{Use}} = \frac{I_{Stress}}{I_{Use}} \tag{14.41}$$

Inserting the expression for the corrosion current yields Peck's equation (Section 9.6)

$$A_{TH} = A_T A_H \tag{14.42}$$

where

$$A_T = Exp\left\{\frac{E_a}{K_B}\left[\frac{1}{T_{Use}} - \frac{1}{T_{Stress}}\right]\right\} \tag{14.43}$$

Here $\Delta G_f/R = E_a/K_B$ and

$$A_H = \left(\frac{RH_{Stress}}{RH_{Use}}\right)^m \tag{14.44}$$

In this case, RH has replaced rh in the ratio to be consistent with Peck's acceleration model in Chapter 9, Section 9.6. Here, m is positive, so that $A_H > 1$. That is, microelectronic failure due to corrosion is accelerated under higher humidity conditions than normally occur during use.

14.6 Diffusion

Diffusion often occurs in mass transport processes. In terms of corrosion, this refers to mass transported in the electrolyte solution to and from the electrodes. Mass transport occurs essentially from three processes: (1) convection and stirring, (2) electrical migration due to a field, and (3) diffusion from a concentration gradient. The first two are categorized as being under the control of an external force. Of these three processes, diffusion is the most important. Diffusion in corrosion can be a rate-controlling step. This is often the case in "hot corrosion" or "aqueous corrosion" due to oxidation. Furthermore, many

Figure 14.10

Diffusion concepts

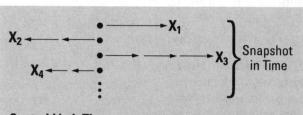

Central Limit Theorem

If we have $X_1, X_2, ..., X_n$ identically distributed independent random variables, each with a mean and a variance, then the normal distrubution as $N \rightarrow$ infinity

aging processes due to diffusion do not involve electrochemical transitions. Diffusion can be understood using a mathematical approach. Consider particles, say impurities; these impurities distribute themselves in space with passing time. For example, in semiconductors, impurities deposited in optimal regions in space later diffuse to undesirable regions as the semiconductor ages. Raising the temperature may accelerate aging.

To describe diffusion mathematically, the Central Limit Theorem (Chapter 8, Figure 8.21) is sometimes useful. For example, the theorem applies for systems subject to a large number of small independent random effects as in a *random walk* (see Reference 6). Here, impurity particles are concentrated in a small region, each with an irregular random walk motion. From the Central Limit Theorem, the positions will become normally distributed in space for times short on a macroscopic scale but long on a microscopic scale. In one dimension, the distribution after time, t, will appear to be Gaussian. Therefore, the probability, P, for finding a particle a distance, x, from the point of initial highest concentration taken as the origin, where one can center the mean (see Figure 8.12), is

$$P(x) = \frac{1}{\sigma\sqrt{2\pi}} \; exp\left[-\frac{1}{2}\left(\frac{x}{\sigma}\right)^2\right] \qquad (14.45)$$

In diffusion theory, for this typical physical situation, it is found that the particles spread linearly with time, t, with the variance as

$$\sigma^2 \propto t \qquad (14.46)$$

The proportionality constant is the diffusion coefficient, D, times 2 (see Reference 6)

$$\sigma^2 = 2Dt \qquad (14.47)$$

(Note that σ has the same units of meters and D has units of square meters per second.) The diffusion coefficient itself is found to have Arrhenius temperature dependence

$$D(T) = D_o \, exp\left\{-\frac{\Delta}{RT}\right\} \qquad (14.48)$$

where Δ is the barrier height (see Section 14.2.2). The probability for finding a particle at position, x, from the origin at time, t, in one-dimension is then

$$P(x,t,T) = \frac{1}{\sqrt{4\pi \; D \; t}} \; exp\left(-\frac{x^2}{4 \; D \; t}\right) \qquad (14.49)$$

In terms of our semiconductor problem, if Q were the number of impurity particles in a unit area and C is concentration of these impurities in the volume, the concentration distribution can be written

$$C(x,t) = \frac{Q}{\sqrt{4\pi \; D \; t}} \; exp\left(-\frac{x^2}{4 \; D \; t}\right) \qquad (14.50)$$

As T increases, so does D (with an Arrhenius form). Note that, since D occurs only as a product Dt, the time scale is effectively changed (accelerating time) with an Arrhenius temperature dependence.

The result is the solution to the diffusion equation with the boundary conditions for a physical situation described above. In one dimension, the diffusion equation is

$$\frac{\partial}{\partial t} C(X,t) = D(T) \frac{\partial^2}{\partial X^2} C(X,t) \qquad (14.51)$$

As an exercise, show that the solution above satisfies this diffusion equation. It is important to note that the solution obtained is subject to the correct initial conditions.

The diffusion acceleration factor is according to the diffusion rate ratio

$$A_D = \frac{Diffusion\ Rate_{Stress}}{Diffusion\ Rate_{Use}} = A_T A_x \qquad (14.52)$$

where A_T is given earlier and A_x is the acceleration factor due to spatial concentration variation

$$A_x = \frac{(\frac{\partial^2 C(x,t)}{\partial x^2})_{stress}}{(\frac{\partial^2 C(x,t)}{\partial x^2})_{Use}} \qquad (14.53)$$

▼ **Example 14.7** *Package moisture time constant*

Problem:
An 85°C and 85%RH test is performed on a plastic molded semiconductor device. It is of interest to estimate how long it takes for the moisture to penetrate the mold and reach the die. Estimate this time using the diffusion expression above. Use the experimentally reported values of

$$\frac{\sigma^2}{2} = 0.85L^2$$

where L is the molding compound thickness of 0.05 inches, $D_o = 4.7 \times 10^{-5}$ m^2/sec for moisture penetration into the mold, and $\Phi = 3 \times 10^{26}$ eV/mole.

Solution:
Solving the diffusion expression above with the time-dependent variance for t reads

$$t = \frac{\sigma^2}{2D} = \frac{0.85L^2}{D_O} \exp\left(\frac{\Phi}{RT}\right)$$

This expression is in terms of R, the gas constant. It is instructive and simplest to put the exponential function in terms of Boltzmann's constant. Boltzmann's constant is by definition $K_B = R/N_o$. To put the expression in terms of Boltzmann's constant one would divide through the exponential expression by Avogadro's number ($N_o = 6 \times 10^{26}$ molecules/mole).

$$exp\left(\frac{\Phi/N_o}{R/N_oT}\right) = exp\left(\frac{E_a}{K_BT}\right)$$

Now $E_a = \Phi/N_o = 0.5$ eV. Inserting the numbers into the above expression reads

$$t = \frac{0.85(0.05 inch \times 0.0254\frac{meter}{inch})^2}{4.7x10^{-5}\frac{m^2}{sec}} exp\left(\frac{0.5eV}{8.617x10^{-5}\frac{eV}{^oK}(85 + 273.16^o K)}\right)$$

$$t = 0.029 sec\; exp(16.2) = 314,752\; sec = 87\; hours$$

The most generalized diffusion equation for aging circumstances can include external forces, such as an electric field. For example, if the flux is a charged species and is driven by a force such as a constant electric field, E, then (see Reference 6)

$$\frac{\partial C(x,t)}{\partial t} = D\frac{\partial^2 C(x,t)}{\partial x^2} - \mu E\frac{\partial C(x,t)}{\partial x} \qquad (14.54)$$

Note that the RHS shows that the diffusion equation can describe all three processes that we described as fundamental: a thermally activated Arrhenius process, the existence of a spatial gradation driving diffusion, and an external forced process. All processes are fundamentally driven by the nonequilibrium thermodynamic state. The equation would be extremely difficult to solve if all mechanisms were equally important. However, aging can often be separated into its rate-controlling process.

14.7 Transistor Aging of Key Device Parameters

Understanding degradation of key transistor parameters is essential in addressing transistor reliability issues. What are some of the key aging kinetics that cause transistor degradation? Understanding device reliability issues through physics-of-failure analysis provides insights that often are unobtainable from experimental work alone. The more one is able to understand, the higher is the likelihood that improvements can be made in device design and in testing methods. In this section, key transistor aging parameters are described to help explain their time dependencies, relationships to device junction degradation, and to aid in predicting parameter degradation. We will note in the bipolar case for the common-emitter configuration, that transistor β aging is directly proportional to the fractional change in the base-emitter leakage current. In the FET case, we will see that transconductance aging results from a change in the drain-source resistance and gate leakage. Modeling also helps explain the observed logarithmic aging of key parameters observed in experiments.

14.7.1 Bipolar Parameter Degradation Aging

Generally, there are two common bipolar aging mechanisms: an increase in emitter ohmic contact resistance and an increase in base leakage currents.

Since β is given by I_{ce}/I_{be} in the common emitter configuration, any base leakage degradation in I_{be} will degrade β.

In this section, we describe a model for β degradation (see Reference 15) over time due to leakage. We start by modeling a change in the base current gain for the common emitter configuration as

$$\beta(t) = \beta_O - \left|\Delta\beta(t)\right| \qquad (14.55)$$

where β_o is the initial value (prior to aging) of I_{ce}/I_{be}. The time-dependent function $\Delta\beta(t)$ can be found through the time derivative

$$\dot{\beta}(t) = \frac{d\beta}{dt} = \frac{d}{dt}\left(\frac{I_{ce}}{I_{be}}\right) = \frac{\dot{I}_{ce}}{I_{be}} - \frac{I_{ce}}{I_{be}^2}\dot{I}_{be} \approx \beta_o\left(\frac{\dot{I}_{ce}}{I_{ce}} - \frac{\dot{I}_{be}}{I_{be}}\right) \qquad (14.56)$$

Approximating d/dt by $\Delta/\Delta t$ and noting that Δt is common to both sides of the equation and cancels, then

$$\Delta\beta(t) \approx \beta_o\left(\frac{\Delta I_{ce}(t)}{I_{ce}} - \frac{\Delta I_{be}(t)}{I_{be}}\right) \approx -\beta_o\left(\frac{\Delta I_{be}(t)}{I_{be}}\right) \qquad (14.57)$$

In the above equation, ΔI_{ce} has been set to zero as no change in this parameter is usually observed experimentally. *Thus, our first result is that the change in β is directly proportional to the fractional change in the base-emitter leakage current.*

At this point, base charge storage is discussed in order to develop a useful capacitive model. When a transistor is first turned on, electrons penetrate into the base bulk gradually. They reach the collector only after a certain delay time, τ_d. The collector current then starts to increase, in relation to the current diffusion rate. Concurrent with the increase of the collector current, excess charges build up in the base. As a first approximation, the collector current and excess charge increase in an exponential manner with time constant, τ_b. This transient represents the process of charging a "capacitor" in the simplest of RC circuits shown in Figure 14.11 (see References 12 and 13). We use this approximation to provide a simple model for base leakage. The steady-state value of excess charge buildup in base-emitter bulk, Q_k, is then in this view

$$Q_k = (Q_{be})_k \cong (C_{be}V_{be})_k = (C_{be}(I_{be}R_{be}))_k = (I_{be}\tau_b)_k \qquad (14.58)$$

where $\tau_b = R_{be}C_{be}$ is the time constant for steady-state excess charges in the base-emitter junction ($\tau_b \gg \tau_d$). As discussed above, the base-emitter junction primarily contributes to aging effects. Along with this bulk effect is parasitic surface charging, Q_s, and leakage. We can also treat these using a simple RC charging model. In this view, the surface leakage can be expressed as

$$Q_s = (Q_{be})_s \cong (C_{be}V_{be})_s = (C_{be}(I_{be}R_{be}))_s = (I_{be}\tau_b)_s \qquad (14.59)$$

The total excess charging at the base is

$$Q_{be} = Q_s + Q_k \qquad (14.60)$$

As the transistor ages, Q_{be} increases along with I_b. Some of the increase in Q_{be} is caused by the increase in impurities and defects in the base surface and bulk regions due to operating stress.

Figure 14.11

Capacitive leakage model

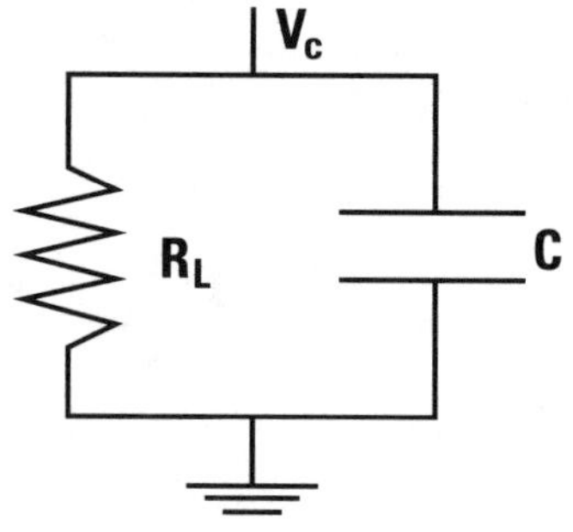

The impurities and defects cause an increase in electron scattering and an increase in the probability for trapping and charging and eventual recombination in the base. The above features lead to an increased leakage current. In the capacitive model shown in Figure 14.11, incremental changes are

$$dQ = C\, dV = C\, R\, dI = \tau\, dI \qquad (14.61)$$

Here, we view Q, V, and I as time varying with age, i.e.,

$$\Delta\beta(t) \cong -\beta_o\left(\frac{\Delta I_{be}(t)}{I_{be}}\right) = -\beta_o\left(\frac{\Delta Q_{be}(t)}{Q_{be}}\right) = -\beta_o\left(\frac{\Delta V_{be}(t)}{V_{be}}\right) \qquad (14.62)$$

Thus, our second result is that the change in β is proportional to the fractional change in the base-emitter leakage current, charge, and voltage.

Experimentally, β degradation follows the aging equation (see Example 14.8)

$$\frac{\Delta\beta(t)}{\beta_o} = \frac{\Delta I_{be}(t)}{I_{be}} = A\, Log(1 + B\, t) \qquad (14.63)$$

In Section 14.8, we will provide an explanation to this observed logarithmic-with-time aging that is consistent with the modeling results and observed FET phenomena discussed next.

14.7.2 FET Parameter Degradation

In this section, transconductance degradation over time is described to help understand aging in FET devices, such as MESFETs, and what role problems like leakage current play (see Reference 15). We start by modeling a change in the transconductance, g_m, as a function of time as

$$g_m(t) = g_o - \left|\Delta g_m(t)\right| \qquad (14.64)$$

where g_o, the initial value taken in the linear portion of the transconductance curve, is

$$g_o = \frac{I_{DS}}{V_{GS} - V_o} \qquad (14.65)$$

Here, we use the linear portion of the curve for simplicity. Similar results will follow for other portions of the curve. The time-dependent function $\Delta g_m(t)$ is found from its derivative as

$$\dot{g}_m(t) = \frac{dg_m}{dt} = \frac{d}{dt}\left(\frac{I_{DS}}{V_{GS} - V_o}\right) = \frac{\dot{I}_{DS}}{V_{GS} - V_o} - \frac{I_{DS}}{(V_{GS} - V)^2}\dot{V}_{GS} \qquad (14.66)$$

or

$$\dot{g}_m(t) = \frac{I_{DS}}{V_{GS} - V_o}\left|\frac{\dot{I}_{DS}}{I_{DS}} - \frac{\dot{V}_{GS}}{V_{GS} - V_o}\right| = g_o\left(\frac{\dot{I}_{DS}}{I_{DS}} - \frac{\dot{V}_{GS}}{V_{GS} - V_o}\right) \qquad (14.67)$$

We assume that the drain-source current change occurs as $dI/dt \sim (V/R^2)(dR/dt)$ with V_{DS} constant and voltage-gate change as $dV_{GS}/dt \sim d/dt(IR) = R\,dI_{GS}/dt$. Approximating d/dt by $\Delta/\Delta t$ and noting that Δt is common to both sides of the equation and cancels, the expression simplifies to

$$\Delta g_m(t) = g_o\left(\frac{\Delta R_{DS}}{R_{DS}} - \frac{\Delta I_{GS}}{I_{GS} - I_{GS_o}}\right) \qquad (14.68)$$

Thus, our primary result for FETs is that transconductance aging arises from a change in the drain-source resistance and gate leakage. However, it is commonly found that resistance aging dominates the reaction (see Reference 12). As far as R_{DS} is concerned, *resistance is related to scattering inside the drain-source channel* $\Delta R_{DS}/R_{DS} = \Delta \rho_{DS}/\rho_{DS} = \Delta l_{DS}/l_{DS}$, where ρ is the resistivity, and l is the average mean-free path the electrons in the channel travel between collisions. This distance decreases as aging occurs, and more defects occur in the channel, causing increased scattering.

At this point, we wish to point out that similar to β degradation (a mechanism that we have modeled as dominated by leakage), MESFET gate leakage data as shown commonly follow a logarithmic-in-time aging form as well (see Example 14.8)

$$\frac{\Delta I_{GS}(t)}{I_{GS}} = A\,Log(1 + B\,t) \qquad (14.69)$$

In the next section, the relevance to the leakage mechanism is discussed.

14.8 Understanding Logarithmic-in-Time Parametric Aging Associated with Activated Processes

When activation is the rate-controlling process, Arrhenius-type rate kinetics applies. In this section, the parametric time-dependence of an Arrhenius mechanism is addressed. This mechanism turns out to explain logarithmic-in-time aging of many key device parameters such as leakage degradation in transistors. Such a mechanism, with temperature as the fundamental thermodynamic stress factor, leads to this predictable logarithmic-in-time-dependent aging on measurable parameters.

There are two Thermally Activated Time-dependent (TAT) models [7 and 8] used to describe the time-dependence of the Arrhenius aging processes. Both will be described.

In Arrhenius processes, the probability rate, dp/dt, to surmount the relative minimum free energy barrier, ϕ (see Section 14.2.2), is

$$\frac{dp}{dt} = v\,exp\left(-\frac{\phi}{K_B T}\right) \qquad (14.70)$$

where v = a rate constant, K_B = Boltzmann's constant, T = the temperature, and t = time. We wish to associate the thermodynamic aging kinetics in the material with measurable parametric changes. Therefore, we model the above as a fractional rate of parametric change given by

$$\frac{da}{dt} = v\,exp\left(-\frac{\phi}{K_B T}\right) \qquad (14.71)$$

where a is a unitless fractional change of the measurable parameter, P (e.g., $a = \Delta P/P_o$). For example, ΔP could be a parameter change that is of concern, such as resistance change, current change, mechanical creep strain change, voltage transistor gain change, and so forth, such that a is then the fractional resistance, current, etc., change (possibly in parts per million).

This model can only be a function of the fractional change if the aging process is closely related to the parametric change. This implies that the free energy itself will be associated with the parameter through the thermodynamic work. Thus, ϕ will be a function of a. This is the basic assumption of the TAT model. Then the free energy can be expanded in terms of its parametric dependence using a Maclaurin series (with environmental factors, held constant for the moment). The free energy reads

$$\phi(a) = \phi(0) + ay_1 + \frac{a^2}{2}y_2 + \dots \qquad (14.72)$$

where y_1 and y_2 are given by

$$y_1 == \frac{\partial \phi(0)}{\partial a} \quad and \quad y_2 = \frac{\partial^2 \phi(0)}{\partial a^2} \qquad (14.73)$$

14.8.1 Arrhenius Aging Due to Small Parametric Change

When a<<1, the first and second terms in the Maclaurin series yield

$$\frac{da}{dt} = v(T) \exp\left(-\frac{ay1}{K_B T}\right) \qquad (14.74)$$

where

$$v(T) = v_o \exp\left(-\frac{\varphi(0)}{K_B T}\right) \qquad (14.75)$$

Rearranging terms, and solving for a as a function of t and integrating, provides a logarithmic-in-time aging TAT model where

$$a = \frac{\Delta P}{P} \cong A \ln[1 + B t] \qquad \text{For } a << 1 \qquad (14.76)$$

Here A and B are

$$A = \frac{K_B T}{y_1} \quad and \quad B = \frac{v(T)y1}{K_B T} \qquad (14.77)$$

Logarithmic-in-time aging is an extremely important process in TRE since the origin of this aging kinetics can mathematically be tied to the Arrhenius mechanisms of which numerous experimental examples exist and are given in References 7 and 8. Figure 14.12 illustrates typical logarithmic-in-time aging. One notes that aging is highly nonlinear for small time. This curve is representative of many aging and kinetic processes such as crystal frequency aging (see References 7, 10, and 11), corrosion of thin films, gate oxide stressing, chemisorption processes, early degradation of primary battery life (see Reference 4), creep (see Reference 9), cold-worked metal recrystallization, superconducting ring flux leakage (yielding degradation of current in the ring), 1/f noise [note that 1/f transforms in the time-domain to log(time)], and so forth. The significance of parametric logarithmic-in-time aging can further be put in perspective as it can be tied to catastrophic log-normal failure rates of semiconductors. This is discussed in References 7 and 8.

14.8.2 Parametric Aging at End of Life Due to the Arrhenius Mechanism

A second TAT model can be obtained for both the initial aging period and end-of-life using both terms in the Maclaurin expansion above and performing the integration. The results obtained in Reference 7 are

$$a = \xi + b \ \ erf^{-1}[\exp(-K^2)(\beta \frac{v(T)}{b})t + erf(K)] \qquad (14.78)$$

where erf and erf^{-1} are the error function and its inverse, and

$$\xi = y1 / y2, \quad b = -2K_BT / y2, \quad K = \xi / b, \quad and \quad \beta = 2 / \sqrt{2\pi} \qquad (14.79)$$

This model is a parametric aging phenomenon that ages similar to logarithmic-in-time models and quickly goes catastrophic at end-of-life due to Arrhenius degradation. This is illustrated in Figure 14.13. The figure shows that aging starts off similar to logarithmic-in-time aging and then quickly goes catastrophic at the critical value a_c corresponding to a critical time t_c. The reader is referred to References 7 or 8 for details.

Figure 14.13 illustrates a number of rate processes. Some examples are batteries (see Reference 7), the three phases of creep (see Reference 9), and cold-worked metals recrystallizing exhibit forms of this dependence over time (see References 7 and 8 for more details). What is interesting in this model is that the rate of initial aging is mathematically connected to its rate of final catastrophic behavior. This suggests that if the initial aging process is truly understood, a catastrophic prognostic is possible!

Figure 14.12
Logarithmic-in-time aging for small fractional change of a over time t

Small Change
$\Delta P/P_o = ALog (1 + B \ Time)$

Time

14.8.3 Modeling the Activation Free Energy

The activation free energy in an Arrhenius process has been previously described as a roller-coaster path (see Section 14.2.2) with a relative minimum. First, we will provide a model and then illustrate it with an example. Figure 14.14 depicts the free energy path with a relative minimum in

Figure 14.13
Aging over all time t with critical values a_c and t_c occurring prior to catastrophic failure

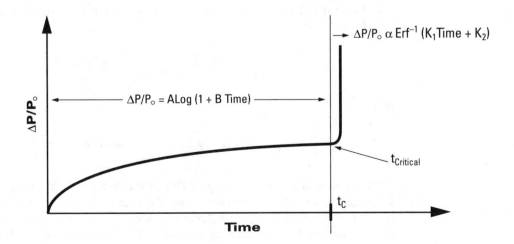

$\Delta P/P_o \ \alpha \ Erf^{-1} (K_1 Time + K_2)$

$\Delta P/P_o$

$\Delta P/P_o = ALog (1 + B \ Time)$

$t_{Critical}$

t_c

Time

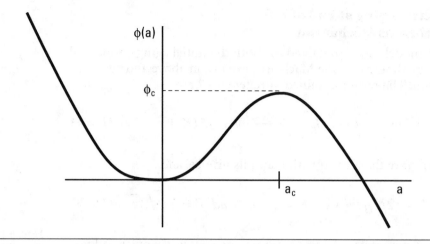

Figure 14.14
Arrhenius activation free energy path having a relative minimum as a function of generalized parameter a

Arrhenius processes. We have centered the axis about the local minimum.

A simple parabolic expression can be used to model the activation free energy near a_c. For example,

$$\phi(a) = \phi_c \{1 - (1 - \frac{a}{a_c})^2\} \qquad (14.80)$$

In terms of the generalized coordinates of force, $f(x)$, and displacement, x, the isothermal work required to change the free energy to its barrier height, ϕ_c, is

$$W = \int_{x_o}^{x_c} F(x)dx = \phi_c \qquad (14.81)$$

From the parabolic free energy model, the Maclaurin expansion first and second derivative terms above for the free energy with respect to a are

$$\frac{\partial \phi(a)}{\partial a} = \frac{2\phi_c}{a_c} \{1 - \frac{a}{a_c}\}, \quad \frac{\partial^2 \phi(a)}{\partial a^2} = -\frac{2\phi_c}{a_c^2} \qquad (14.82)$$

At $a = 0$, the first two partial derivative terms in the expansion are then

$$y1 = \frac{\partial \phi(a = 0)}{\partial a} = \frac{2\phi_c}{a_c}, \quad y2 = \frac{\partial^2 \phi(a)}{\partial a^2} = -\frac{2\phi_c}{a_c^2} = \frac{y1}{a_c} \qquad (14.83)$$

For example, in the case where $a \ll 1$, the logarithmic-in-time aging solution from above is

$$a = \frac{\Delta x}{x_O} \cong \frac{K_B T}{y1} \ \ln[1 + \frac{v(T)y1}{K_B T} \ t] \qquad \text{For } a \ll 1 \qquad (14.84)$$

where

$$\Delta x_c = x_c - x_o, \ \Delta x \equiv x - x_o, \quad a \equiv \frac{\Delta x}{x_O} \ \text{and} \ a_c \equiv \frac{\Delta x_c}{x_O} \qquad (14.85)$$

We are now in a position to apply this model to a practical problem of interest involving any thermodynamic state variable such as those for the generalized coordinates in Table 14.1. Depending on the physical aging mechanism that is modeled by Arrhenius rate kinetics, the TAT model provides a reasonable description of aging kinetics. Usually, ϕ_c may be experimentally determined in terms of a key parameter, but its dependencies are important to identify as illustrated in the following example.

▼ Example 14.8 *Capacitance leakage*

Problem:

Capacitance leakage is an important microelectronic problem. For example, β degradation in bipolar transistors and transconductance degradation in FETs, when dominated by leakage current problems as discussed in Section 14.7, may be viewed as a capacitor leakage problem. As an example, the TAT model is used to provide an aging model for capacitance charge leakage as modeled in Figure 14.10 in terms of its design parameter C, V, R, and a critical leakage charge q_c.

Solution:

In this case, the mechanical variables from Table 14.1 are $dW = Vdq$. Noting that $V = q/C$ (C is the initial capacitance), the work required to change the free energy to its critical value is

$$W = \int_{q_o}^{q_c} \frac{q}{C} dq = \frac{\Delta q_c^2}{2C} = \phi_c \qquad (14.86)$$

Noting that

$$\Delta q_c = q_c - q_o, \quad \Delta q \equiv q - q_o, \quad a \equiv \frac{\Delta q}{q_o} \quad \text{and} \quad a_c \equiv \frac{q - q_c}{q_o} \qquad (14.87)$$

the free energy is

$$\phi(a) = \frac{\Delta q_c^2}{2C} \{1 - (-\frac{a}{a_c})^2\} \qquad (14.88)$$

The key parameters $y1$ and $y2$ are

$$y1 = \frac{2\phi_c}{a_c} = \frac{\Delta q_c q_o}{C}, \quad y2 = \frac{\partial^2 \phi(a)}{\partial a^2} = -\frac{y1}{a_c} = -\frac{q_o^2}{C} \qquad (14.89)$$

For a small parameter change, the aging equation is

$$a = \frac{\Delta q}{q_o} \cong \frac{CK_B T}{\Delta q_c q_o} \ln[1 + \frac{v(T)\Delta q_c q_o}{CK_B T} t] \quad \text{For } a \ll 1 \qquad (14.90)$$

or in terms of the leakage itself

$$\Delta q \cong \frac{CK_B T}{\Delta q_c} \ln[1 + \frac{v(T)\Delta q_c q_o}{CK_B T} t] \qquad (14.91)$$

The model shows the logarithmic-in-time aging dependencies. Note that the leading term indicates a linear dependence on aging with the capacitance. This indicates that the leakage will be directly proportional to the design parameters of C (i.e., $C = \kappa A/d$).

Using the leakage current model in Figure 14.11

$$\Delta I_{leakage} = \Delta V/R_L \tag{14.92}$$

which is proportional to the effective capacitance charge ($\Delta V = \Delta q/C$) and the leakage $\Delta I = \Delta q/CRL$. Thus, a can be written in terms of the fractional leakage current

$$a = \frac{\Delta I}{I_o} \cong \frac{CK_BT}{\Delta q_c q_o} ln[1 + \frac{v(T)\Delta q_c q_o}{CK_BT}t] \tag{14.93}$$

The K_BT/y_1 term can be equated to

$$\frac{CK_BT}{q_o\Delta q_c} = \frac{K_BT}{V_oC\Delta V_c} = \frac{K_BT}{I_o(RC)\Delta V_c} = \frac{K_BT}{\tau I_o\Delta V_c} \tag{14.94}$$

where $\tau = R_LC$. Using the last expression, the leakage current is

$$\Delta I = \left(\frac{K_BT}{\tau\Delta V_c}\right)ln\left[1 + \left(\frac{\tau I_o v(T)\Delta V_c}{K_BT}\right)t\right] \tag{14.95}$$

This logarithmic-in-time form for leakage current has been clearly observed experimentally (see Reference 12) in both bipolar and FET life test data. For example, from Section 14.7.1, the bipolar β A and B parameters are identified by comparison

$$\frac{\Delta\beta(t)}{\beta_o} = \frac{\Delta I_{be}(t)}{Ibe} = A\,Log(1 + B\,t) \tag{14.96}$$

and similarly, in the FET case for gate leakage.

Thus, the central finding is the logarithmic-in-time aging dependencies. For example, the leading term in front of the mathematical log indicates that leakage is proportional to temperature and inversely proportional to the capacitor charging time constant and the critical voltage. The critical (or breakdown) voltage is an intrinsic quantity. The model indicates that design parameters of breakdown voltage and $\tau = R_LC$ values can help control leakage associated with aging.

The catastrophic model provides the dependencies for all time including the catastrophic critical points (see Reference 7)

$$a = -a_c + b\;erf^{-1}[\exp(-K^2)(\beta\frac{v(T)}{b})t + erf(K)] \tag{14.97}$$

where

$$b = \sqrt{\frac{2CKT}{q_o^2}}, \quad K = \frac{a_c}{b} \tag{14.98}$$

14.9 Summary

This chapter provides an overview of the relationship between reliability physics and the science of thermodynamics. We have linked these by providing TRE principles. A number of examples have been provided including a derivation of Miner's rule and time-compression Coffin-Manson temperature cycle, Peck humidity, and a vibration model. The results demonstrate the usefulness of applying powerful thermodynamic tools to obtain physics-of-failure models.

References

1. Feinberg, A. A., and Widom, A., "On Thermodynamic Reliability Engineering," submitted to *IEEE Transactions on Reliability*, (this reference may be helpful to augment this material) 1999.

2. Feinberg, A. A., and Widom, A., "Aspects of the Thermodynamic Aging Process in Reliability Physics," *Institute of Environmental Sciences Proceedings*, p. 49, May 1995.

3. Miner, "Cumulative Damage in Fatigue," *Journal of Applied Mechanics* 12, A159-A164, 1945.

4. Linden, D., Editor in Chief, *Handbook of Batteries and Fuel Cells*, McGraw-Hill, New York, 1980, pp. 1-21, 14-9, 14-38, 14-39, 14-59, 14-79.

5. Fontana, M. G., and Greene, N. D., *Corrosion Engineering*, McGraw-Hill, New York, 1978, p. 356.

6. Morse, *Thermal Physics*, Benjamin/Cummings Pub. Co.

7. Feinberg, A. A., and Widom, A., "Connecting Parametric Aging to Catastrophic Failure Through Thermodynamics," *IEEE Transactions on Reliability*, Vol. 45, No. 1, p. 28, March 1996.

8. Feinberg, A. A., and Widom, A., "The Reliability Physics of Thermodynamic Aging," *Recent Advances in Life-Testing and Reliability*, Edited by N. Balakrishnan, CRC Press, April 1995.

9. Feinberg, A. A., "Modeling Creep Using a Thermodynamic TAT Model in Reliability Physics," *Institute of Environmental Sciences Proceedings*, p. 50, May 1996.

10. Feinberg, A. A., "Gaussian Parametric Failure-Rate Model with Application to Quartz-Crystal Device Aging," *IEEE Transactions on Reliability*, p. 565, 1992.

11. Warner, A. W., Fraser, D. B., and Stockbridge, C. D., "Fundamental Studies of Aging in Quartz Resonators," *IEEE Transactions on Sonics and Ultrasonics*, p. 52, 1965.

12. Feinberg, A. A., Ersland, P., and Kaper, V., "Modeling and Understanding Junction Temperature-Dependent Leakage Degradation," AMP M/A-COM Engineering Conference, pp. 197-202, Oct. 1998. Also submitted to *IEEE Transactions on Electron Devices*, 1999.

13. Stepanenko, I. P., *Fundamentals of Microelectronics*, MIR Publishers, Moscow, 1983.

14. Feinberg, A. A., and Widom, A., "Thermodynamic Extensions of Miner's Rule to Chemical Cells," *Reliability and Maintainability Symposium*, p. 341, 2000.

15. Feinberg, A. A., Ersland, P., Kaper, V., and Widom, A., "On Transistor Aging of Key Device Parameters," *Institute of Environmental Sciences and Technology*, 2000.

SUBJECT INDEX